Essentials of Hydrology

Essentials of Hydrology

Editor: Shaun Grantham

R CALLISTO REFERENCE

www.callistoreference.com

Callisto Reference,
118-35 Queens Blvd., Suite 400,
Forest Hills, NY 11375, USA

Visit us on the World Wide Web at:
www.callistoreference.com

ISBN: 978-1-64116-013-1 (Hardback)

Cataloging-in-Publication Data

Essentials of hydrology / edited by Shaun Grantham.
    p. cm.
Includes bibliographical references and index.
ISBN 978-1-64116-013-1
1. Hydrology. 2. Ecohydrology. I. Grantham, Shaun.
GB661.2 .E87 2018
551.48--dc23

Table of Contents

Preface

The study of the quality, movement, distribution, properties and characteristics of water present on Earth as well as other planets is known as hydrology. It also includes topics like environmental watershed management, water cycle and water resources management. The field has many sub-branches like ecohydrology, isotope hydrology, water quality, chemical hydrology, hydroinformatics, hydrogeology, drainage basin assessment, surface hydrology, hydrometeorology, etc. This textbook is a valuable compilation of topics, ranging from the basic to the most complex theories and principles in the field of hydrology. It will prove immensely beneficial to students involved in this area at various levels.

To facilitate a deeper understanding of the contents of this book a short introduction of every chapter is written below:

Chapter 1- Hydrologic or water cycle forms the basic component of hydrology. The movement of water in all its phases from the Earth to the atmosphere and back is called water cycle. Hydrology can be defined as the study of the movement, quality, occurrence and distribution of water. The chapter on hydrologic cycle and its measurement offers an insightful focus, keeping in mind the complex subject matter.

Chapter 2- A continuous graph of rate of flow versus time is known as hydrograph. Analyzing hydrography is essential for flood forecasting, flood damage reduction, and creating design flows for structures to channel floodwater. Influence of hydrography's shape and volume are affected by factors such as storm event's duration, soil types and distribution, rainfall intensity and pattern, etc. The diverse applications of hydrograph in the current scenario have been thoroughly discussed in this chapter.

Chapter 3- Statistical analysis is preferred to observing hydrological variables by applying physical laws. The reason is that the data received is inadequate, plausibly incorrect, and inherits randomness. Statistics also allows for probability and adjustability to real-world situations. This chapter is an overview of the subject matter incorporating all the major aspects of hydrology.

Chapter 4- The connection between hydrologic variables can best be understood through cause and effect. If a change occurs in one variable, the other variable also changes. Regression analysis is the statistical method of measuring the relationship between variables. The topics discussed in the section are of great importance to broaden the existing knowledge on the subject matter.

Chapter 5- A hydrologic model represents a hydrologic cycle. Hydrologic models can be categorized into two models, stochastic models and process-based models. Hydrological Simulation Program-Fortran (HSPF), MIKE, etc. are some of the hydrologic models explored in this chapter. The aspects elucidated in this chapter are of vital importance, and provide a better understanding of hydrologic models.

Finally, I would like to thank the entire team involved in the inception of this book for their valuable time and contribution. This book would not have been possible without their efforts. I would also like to thank my friends and family for their constant support.

Editor

Basics of Hydrologic Cycle

Hydrologic or water cycle forms the basic component of hydrology. The movement of water in all its phases from the Earth to the atmosphere and back is called water cycle. Hydrology can be defined as the study of the movement, quality, occurrence and distribution of water. The chapter on hydrologic cycle and its measurement offers an insightful focus, keeping in mind the complex subject matter.

Hydrologic Cycle

Water can occur in three physical phases: solid, liquid, and gas and is found in nature in all these phases in large quantities. Depending upon the environment of the place of occurrence, water can quickly change its phase.

A number of cycles are operating in nature, such as the carbon cycle, the nitrogen cycle, and several biogeochemical cycles. The Hydrologic Cycle, also known as the water cycle, is one such cycle which forms the fundamental concept in hydrology. Hydrologic cycle was defined by the National Research Council (NRC, 1982) the as "the pathway of water as it moves in its various phases to the atmosphere, to the earth, over and through the land, to the ocean and back to the atmosphere". This cycle has no beginning or end and water is present in all the three states (solid, liquid, and gas). A pictorial view of the hydrological cycle is given in figure. The science of hydrology primarily deals with the land portion of the hydrologic cycle; interactions with the oceans and atmosphere are also studied. NRC (1991) called the hydrologic cycle as the integrating process for the fluxes of water, energy, and the chemical elements.

The hydrologic cycle can be visualized as a series of storages and a set of activities that move water among these storages. Among these, oceans are the largest reservoirs, holding about 97% of the earth's water. Of the remaining 3% freshwater, about 78% is stored in ice in Antarctica and Greenland. About 21% of freshwater on the earth is groundwater, stored in sediments and rocks below the surface of the earth. Rivers, streams, and lakes together contain less than 1% of the freshwater on the earth and less than 0.1% of all the water on the earth.

Hydrologic cycle considers the motion, loss, and recharge of the earth's waters. It connects the atmosphere and two storages of the earth system: the oceans, and the land sphere (lithosphere and pedosphere). The water evaporated from the earth and the oceans enters the atmosphere. Water leaves the atmosphere through precipitation. The oceans receive water from the atmosphere by means of precipitation and from the land through rivers and ground water flow. Water goes out of oceans only through evaporation. The water leaves land through evapotranspiration, streamflow, and ground water flow. Evaporation and precipitation processes take place in the vertical plane while ground water flow occur mostly in the streamflow and horizontal plane.

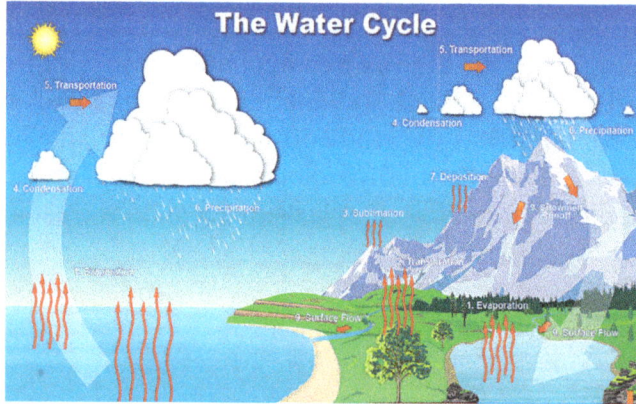

Pictorial view of the hydrological cycle.

The exchange of water among the oceans, land, and the atmosphere was termed as 'the turnover' by Shikhlomanov (1999). This turnover affects the global patterns of the movement of ocean waters and gases in the atmosphere, thereby greatly influencing climate. Since water is a very good solvent, chemistry is an integral part of the hydrologic cycle. Usually, rain and snow are considered as the purest form of water although these may also be mixed with pollutants that are present in the atmosphere. During the journey on earth, many chemical compounds are mixed with water and consequently the water quality undergoes a change. One can also visualize the hydrologic cycle as a perpetual distillation and pumping system in which the glaciers and snow packs are replenished and rivers get water of good quality.

We need to study the hydrologic cycle since water is essential for survival of life and is an important input in many economic activities. From the use point of view, the land phase of the hydrologic cycle is the most important.

In view of the complexities and extensive coverage, the study of the complete hydrologic cycle is truly interdisciplinary. For instance, the atmospheric part is studied by meteorologists, the pedospheric part by soil scientists, the lithosphere part by geologists, and the part pertaining to oceans falls in the domain of oceanographers. A host of other professionals study hydrologic cycle: agricultural engineers, energy managers, ecologists and environmentalists, public health officers, industrialists, chemists, and inland navigation managers.

Components of Hydrologic Cycle

The hydrologic cycle can be subdivided into three major systems: The oceans being the major reservoir and source of water, the atmosphere functioning as the carrier and deliverer of water and the land as the user of water. The amount of water available at a particular place changes with time because of changes in the supply and delivery. On a global basis, the water movement is a closed system but on a local basis it is an open system.

The major components of the hydrologic cycle are precipitation (rainfall, snowfall, hale, sleet, fog, dew, drizzle, etc.), interception, depression storage, evaporation, transpiration, infiltration, percolation, moisture storage in the unsaturated zone, and runoff (surface runoff, interflow, and baseflow).

Evaporation of water takes place from the oceans and the land surface mainly due to solar energy. The moisture moves in the atmosphere in the form of water vapour which precipitates on land surface or oceans in the form of rain, snow, hail, sleet, etc. A part of this precipitation is intercepted by vegetation or buildings. Of the amount reaching the land surface, a part infiltrates into the soil and the remaining water runs off the land surface to join streams. These streams finally discharge into the ocean. Some of the infiltrated water percolates deep to join groundwater and some comes back to the streams or appears on the surface as springs.

This immense movement of water is mainly driven by solar energy: the excess of incoming radiation over the outgoing radiation. Therefore, sun is the prime mover of the hydrologic cycle. The energy for evaporation of water from streams, lakes, ponds and oceans and other open water bodies comes from sun. A substantial quantity of moisture is added to the atmosphere by transpiration of water from vegetation. Living beings also supply water vapor to the atmosphere through perspiration. Gravity has an important role in the movement of water on the earth's surface and anthroprogenic activities also have an increasingly important influence on the water movement.

An interesting feature of the hydrologic cycle is that at some point in each phase, there usually occur: (a) transportation of water, (b) temporary storage, and (c) change of state. For example, in the atmospheric phase, there occurs vapor flow, vapor storage in the atmosphere and condensation or formation of precipitation created by a change from vapor to either the liquid or solid state. Moreover, in the atmosphere, water is present in the vapor form while it is mostly (saline) liquid in the oceans.

Scales for Study of Hydrologic Cycle

From the point of view of hydrologic studies, two scales are readily distinct. These are the global scale and the catchment scale.

Global Scale

From a global perspective, the hydrologic cycle can be considered to be comprised of three major systems; the oceans, the atmosphere, and the landsphere. Precipitation, runoff and evaporation are the principal processes that transmit water from one system to the other. This depicts a global geophysical view of the hydrologic cycle and shows the interactions between the earth (lithosphere), the oceans (hydrosphere), and the atmosphere. The study at the global scale is necessary to understand the global fluxes and global circulation patterns. The results of these studies form important inputs to water resources planning for a national, regional water resources assessment, weather forecasting, and study of climate changes. These results may also form the boundary conditions of small-scale models/applications.

Catchment Scale

While studying the hydrologic cycle on a catchment scale, the spatial coverage can range from a few square km to thousands of square km. The time scale could be a storm lasting for a few hours to a study spanning many years. When the water movement of the earth system is considered, three systems can be recognized: the land (surface) system, the subsurface system, and the aquifer (or geologic) system. When the attention is focused on the hydrologic cycle of the land system, the dominant processes are precipitation, evapotranspiration, infiltration, and surface runoff. The land system itself comprises of three subsystems: vegetation subsystem, structural subsystem and soil subsystem. These subsystems subtract water from precipitation through interception, depression and detention storage. This water is either lost to the atmospheric system or enters subsurface system. The exchange of water among these subsystems takes place through the processes of infiltration, exfiltration, percolation, and capillary rise.

Figure shows the schematic of the hydrologic cycle at global scale, in the earth system, and micro-scale view of the cycle in the land system. Figure gives a schematic presentation of the hydrologic cycle of the earth system. Detailed schematic of the hydrologic cycle in the land system is shown in figure.

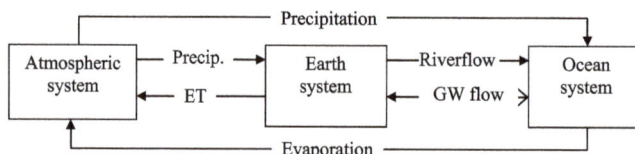

A global schematic of the hydrologic cycle.

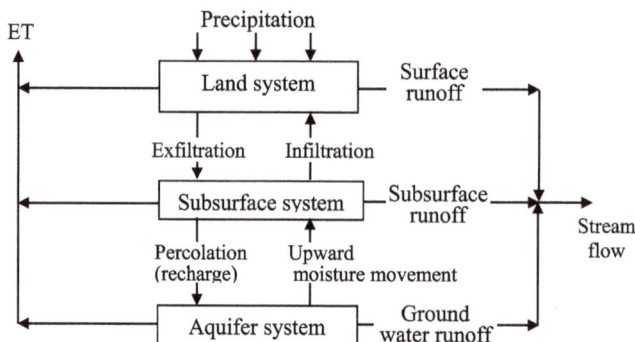

A schematic of the hydrologic cycle of the earth system.

Time Scales in Hydrologic Cycle

The time required for the movement of water through various components of the hydrologic cycle varies considerably. The velocity of streamflow is much higher compared to the velocity of ground water. The time-step size for an analysis depends upon the purpose of study, the availability of data, and how detailed the study is. The estimated periods of renewal of water resources in water bodies on the earth is given in the table. The time step should be sufficiently small so that the variations in the processes can be captured in sufficient detail but at the same time, it should not put undue burden on data collection and computational efforts.

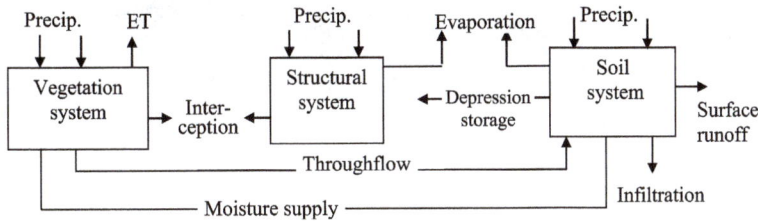

A detailed schematic of the hydrologic cycle in the land system.

Table: Periods of water resources renewal on the Earth

Water of hydrosphere	Period of renewal
World Ocean	2500 years
Ground water	1400 years
Polar ice	9700 years
Mountain glaciers	1600 years
Ground ice of the permafrost zone	10000 years
Lakes	17 years
Bogs	5 years
Soil moisture	1 year
Channel network	16 days
Atmospheric moisture	8 days
Biological water	Several hours

Hydrology

Hydrology is the scientific study of the movement, distribution, and quality of water on Earth and other planets, including the water cycle, water resources and environmental watershed sustainability. A practitioner of hydrology is a hydrologist, working within the fields of earth or environmental science, physical geography, geology or civil and environmental engineering. Using various analytical methods and scientific techniques, they collect and analyze data to help solve water related problems such as environmental preservation, natural disasters, and water management.

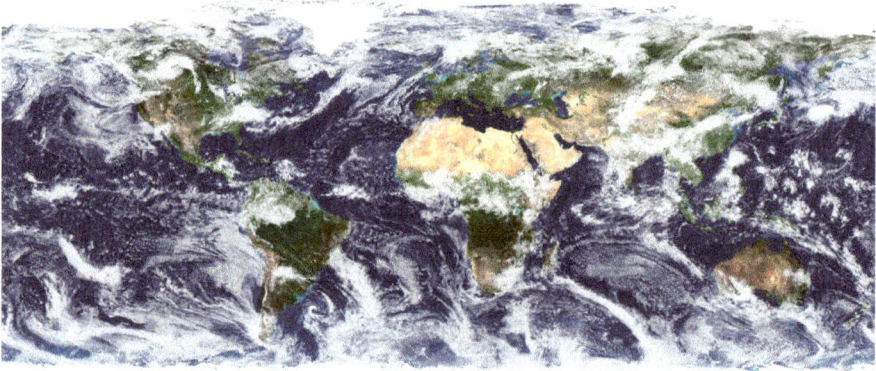

Water covers 70% of the Earth's surface.

Hydrology sub-divides into surface water hydrology, groundwater hydrology (hydrogeology), and marine hydrology. Domains of hydrology include hydrometeorology, surface hydrology, hydrogeology, drainage-basin management and water quality, where water plays the central role.

Oceanography and meteorology are not included because water is only one of many important aspects within those fields.

Hydrological research can inform environmental engineering, policy and planning.

Branches

- Chemical hydrology is the study of the chemical characteristics of water.

- Ecohydrology is the study of interactions between organisms and the hydrologic cycle.

- Hydrogeology is the study of the presence and movement of groundwater.

- Hydroinformatics is the adaptation of information technology to hydrology and water resources applications.

- Hydrometeorology is the study of the transfer of water and energy between land and water body surfaces and the lower atmosphere.

- Isotope hydrology is the study of the isotopic signatures of water.

- Surface hydrology is the study of hydrologic processes that operate at or near Earth's surface.

- Drainage basin management covers water-storage, in the form of reservoirs, and flood-protection.

- Water quality includes the chemistry of water in rivers and lakes, both of pollutants and natural solutes.

Applications

- Determining the water balance of a region.

- Determining the agricultural water balance.

- Designing riparian restoration projects.

- Mitigating and predicting flood, landslide and drought risk.

- Real-time flood forecasting and flood warning.

- Designing irrigation schemes and managing agricultural productivity.

- Part of the hazard module in catastrophe modeling.

- Providing drinking water.

- Designing dams for water supply or hydroelectric power generation.

- Designing bridges.

- Designing sewers and urban drainage system.

- Analyzing the impacts of antecedent moisture on sanitary sewer systems.

- Predicting geomorphologic changes, such as erosion or sedimentation.

- Assessing the impacts of natural and anthropogenic environmental change on water resources.

- Assessing contaminant transport risk and establishing environmental policy guidelines.

History

Hydrology has been a subject of investigation and engineering for millennia. For example, about 4000 BC the Nile was dammed to improve agricultural productivity of previously barren lands. Mesopotamian towns were protected from flooding with high earthen walls. Aqueducts were built by the Greeks and Ancient Romans, while the history of China shows they built irrigation and flood control works. The ancient Sinhalese used hydrology to build complex irrigation works in Sri Lanka, also known for invention of the Valve Pit which allowed construction of large reservoirs, anicuts and canals which still function.

Marcus Vitruvius, in the first century BC, described a philosophical theory of the hydrologic cycle, in which precipitation falling in the mountains infiltrated the Earth's surface and led to streams and springs in the lowlands. With adoption of a more scientific approach, Leonardo da Vinci and Bernard Palissy independently reached an accurate representation of the hydrologic cycle. It was not until the 17th century that hydrologic variables began to be quantified.

Pioneers of the modern science of hydrology include Pierre Perrault, Edme Mariotte and Edmund Halley. By measuring rainfall, runoff, and drainage area, Perrault showed that rainfall was sufficient to account for flow of the Seine. Marriotte combined velocity and river cross-section measurements to obtain discharge, again in the Seine. Halley showed that the evaporation from the Mediterranean Sea was sufficient to account for the outflow of rivers flowing into the sea.

Advances in the 18th century included the Bernoulli piezometer and Bernoulli's equation, by Daniel Bernoulli, and the Pitot tube, by Henri Pitot. The 19th century saw development in groundwater hydrology, including Darcy's law, the Dupuit-Thiem well formula, and Hagen-Poiseuille's capillary flow equation.

Rational analyses began to replace empiricism in the 20th century, while governmental agencies began their own hydrological research programs. Of particular importance were Leroy Sherman's unit hydrograph, the infiltration theory of Robert E. Horton, and C.V. Theis's aquifer test/equation describing well hydraulics.

Since the 1950s, hydrology has been approached with a more theoretical basis than in the past, facilitated by advances in the physical understanding of hydrological processes and by the advent of computers and especially geographic information systems (GIS).

Themes

The central theme of hydrology is that water circulates throughout the Earth through different pathways and at different rates. The most vivid image of this is in the evaporation of water from the ocean, which forms clouds. These clouds drift over the land and produce rain. The rainwater flows into lakes, rivers, or aquifers. The water in lakes, rivers, and aquifers then either evaporates back to the atmosphere or eventually flows back to the ocean, completing a cycle. Water changes its state of being several times throughout this cycle.

The areas of research within hydrology concern the movement of water between its various states, or within a given state, or simply quantifying the amounts in these states in a given region. Parts of hydrology concern developing methods for directly measuring these flows or amounts of water, while others concern modelling these processes either for scientific knowledge or for making prediction in practical applications.

Groundwater

Ground water is water beneath Earth's surface, often pumped for drinking water. Groundwater hydrology (hydrogeology) considers quantifying groundwater flow and solute transport. Problems in describing the saturated zone include the characterization of aquifers in terms of flow direction, groundwater pressure and, by inference, groundwater depth. Measurements here can be made using a piezometer. Aquifers are also described in terms of hydraulic conductivity, storativity and transmissivity. There are a number of geophysical methods for characterising aquifers. There are also problems in characterising the vadose zone (unsaturated zone).

Infiltration

Infiltration is the process by which water enters the soil. Some of the water is absorbed, and the rest percolates down to the water table. The infiltration capacity, the maximum rate at which the soil can absorb water, depends on several factors. The layer that is already saturated provides a resistance that is proportional to its thickness, while that plus the depth of water above the soil provides the driving force (hydraulic head). Dry soil can allow rapid infiltration by capillary action; this force diminishes as the soil becomes wet. Compaction reduces the porosity and the pore sizes. Surface cover increases capacity by retarding runoff, reducing compaction and other processes. Higher temperatures reduce viscosity, increasing infiltration.

Soil Moisture

Soil moisture can be measured in various ways; by capacitance probe, time domain reflectometer or Tensiometer. Other methods include solute sampling and geophysical methods.

Surface Water Flow

Hydrology considers quantifying surface water flow and solute transport, although the treatment of flows in large rivers is sometimes considered as a distinct topic of hydraulics or hydrodynamics. Surface water flow can include flow both in recognizable river channels and otherwise. Methods for measuring flow once water has reached a river include the stream gauge, and tracer techniques. Other topics include chemical transport as part of surface water, sediment transport and erosion.

One of the important areas of hydrology is the interchange between rivers and aquifers. Groundwater/surface water interactions in streams and aquifers can be complex and the direction of net water flux (into surface water or into the aquifer) may vary spatially along a stream channel and over time at any particular location, depending on the relationship between stream stage and groundwater levels.

Precipitation and Evaporation

In some considerations, hydrology is thought of as starting at the land-atmosphere boundary and so it is important to have adequate knowledge of both precipitation and evaporation. Precipitation can be measured in various ways: disdrometer for precipitation characteristics at a fine time scale; radar for cloud properties, rain rate estimation, hail and snow detection; rain gauge for routine accurate measurements of rain and snowfall; satellite for rainy area identification, rain rate estimation, land-cover/land-use, and soil moisture, for example, evaporation is an important part of the water cycle. It is partly affected by humidity, which can be measured by a sling psychrometer. It is also affected by the presence of snow, hail and ice and can relate to dew, mist and fog. Hydrology considers evaporation of various forms: from water surfaces; as transpiration from plant surfaces in natural and agronomic ecosystems. A direct measurement of evaporation can be obtained using Simon's evaporation pan.

Detailed studies of evaporation involve boundary layer considerations as well as momentum, heat flux and energy budgets.

Remote Sensing

Remote sensing of hydrologic processes can provide information on locations where *in situ* sensors may be unavailable or sparse. It also enables observations over large spatial extents. Many of the variables constituting the terrestrial water balance, for example surface water storage, soil moisture, precipitation, evapotranspiration, and snow and ice, are measurable using remote sensing at various spatial-temporal resolutions and accuracies. Sources of remote sensing include land-based sensors, airborne sensors and satellite sensors which can capture microwave, thermal and near-infrared data or use lidar, for example.

Water Quality

In hydrology, studies of water quality concern organic and inorganic compounds, and both dissolved and sediment material. In addition, water quality is affected by the interaction of dissolved oxygen with organic material and various chemical transformations that may take place. Measurements of water quality may involve either in-situ methods, in which analyses take place on-site, often automatically, and laboratory-based analyses and may include microbiological analysis.

Integrating Measurement and Modelling

- Budget analyses

- Parameter estimation

- Scaling in time and space

- Data assimilation

- Quality control of data

Prediction

Observations of hydrologic processes are used to make predictions of the future behaviour of hydrologic systems (water flow, water quality). One of the major current concerns in hydrologic research is "Prediction in Ungauged Basins" (PUB), i.e. in basins where no or only very few data exist.

Statistical Hydrology

By analyzing the statistical properties of hydrologic records, such as rainfall or river flow, hydrologists can estimate future hydrologic phenomena. When making assessments of how often relatively rare events will occur, analyses are made in terms of the return period of such events. Other quantities of interest include the average flow in a river, in a year or by season.

These estimates are important for engineers and economists so that proper risk analysis can be performed to influence investment decisions in future infrastructure and to determine the yield reliability characteristics of water supply systems. Statistical information is utilized to formulate operating rules for large dams forming part of systems which include agricultural, industrial and residential demands.

Modeling

Hydrological models are simplified, conceptual representations of a part of the hydrologic cycle. They are primarily used for hydrological prediction and for understanding hydrological processes, within the general field of scientific modeling. Two major types of hydrological models can be distinguished:

- Models based on data. These models are black box systems, using mathematical and statistical concepts to link a certain input (for instance rainfall) to the model output (for instance runoff). Commonly used techniques are regression, transfer functions, and system identification. The simplest of these models may be linear models, but it is common to deploy non-linear components to represent some general aspects of a catchment's response without going deeply into the real physical processes involved. An example of such an aspect is the well-known behavior that a catchment will respond much more quickly and strongly when it is already wet than when it is dry.

- Models based on process descriptions. These models try to represent the physical processes observed in the real world. Typically, such models contain representations of surface runoff, subsurface flow, evapotranspiration, and channel flow, but they can be far more complicated. These models are known as deterministic hydrology models. Deterministic hydrology models can be subdivided into single-event models and continuous simulation models.

Recent research in hydrological modeling tries to have a more global approach to the understanding of the behavior of hydrologic systems to make better predictions and to face the major challenges in water resources management.

Transport

Water movement is a significant means by which other material, such as soil, gravel, boulders or pollutants, are transported from place to place. Initial input to receiving waters may arise from a point source discharge or a line source or area source, such as surface runoff. Since the 1960s rather complex mathematical models have been developed, facilitated by the availability of high speed computers. The most common pollutant classes analyzed are nutrients, pesticides, total dissolved solids and sediment.

Organizations

Intergovernmental Organizations

- International Hydrological Programme (IHP)

International Research Bodies

- International Water Management Institute (IWMI)

- UNESCO-IHE Institute for Water Education

National Research Bodies

- Centre for Ecology and Hydrology – UK

- Centre for Water Science, Cranfield University, UK

- Eawag – aquatic research, ETH Zürich, Switzerland

- Institute of Hydrology, Albert-Ludwigs-University of Freiburg, Germany

- United States Geological Survey – Water Resources of the United States

- NOAA's National Weather Service – Office of Hydrologic Development, USA

- US Army Corps of Engineers Hydrologic Engineering Center, USA

- Hydrologic Research Center, USA

- NOAA Economics and Social Sciences, USA

- University of Oklahoma Center for Natural Hazards and Disasters Research, USA

- National Hydrology Research Centre, Canada

- National Institute of Hydrology, India

National and International Societies

- Geological Society of America (GSA) – Hydrogeology Division

- American Geophysical Union (AGU) – Hydrology Section

- National Ground Water Association (NGWA)

- American Water Resources Association

- Consortium of Universities for the Advancement of Hydrologic Science, Inc. (CUAHSI)

- International Association of Hydrological Sciences (IAHS)

- Statistics in Hydrology Working Group (subgroup of IAHS)

- German Hydrological Society (DHG: Deutsche Hydrologische Gesellschaft)

- Italian Hydrological Society (SII-IHS)

- Nordic Association for Hydrology

- British Hydrological Society

- Russian Geographical Society (Moscow Center) – Hydrology Commission

- International Association for Environmental Hydrology

- International Association of Hydrogeologists

Basin- and Catchment-wide Overviews

- Connected Waters Initiative, University of New South Wales – Investigating and raising awareness of groundwater and water resource issues in Australia

- Murray Darling Basin Initiative, Department of Environment and Heritage, Australia

Mathematical Representation of the Hydrologic Cycle

The quantities of water going through the various components of the hydrologic cycle can be evaluated by the so-called hydrologic equation, which is a simple spatially-lumped continuity or water budget equation:

$$I - Q = \Delta S$$

where I = inflow of water to a given area during any given time period, Q = outflow of water from the area during the selected time period, and ΔS = change in storage of water in the given area during the time period. If I and Q vary continuously with time, then above equation can be written as

$$d\,S(t)\,/\,dt = I(t) - Q(t)$$

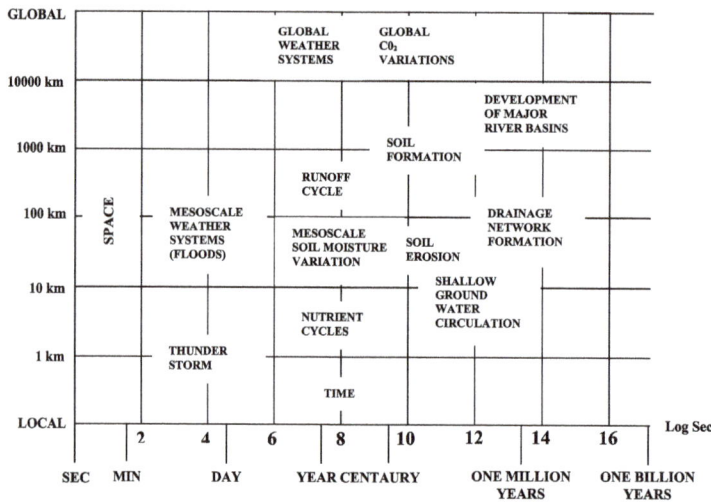

Illustrative range of process scales

By integrating, this equation can also be written as

$$\int dS(t) = \int [I(t) - Q(t)]\,dt$$

or

$$S(t) - S(0) = \int_0^t I(t)dt - \int_0^t Q(t)dt$$
$$= V_I(t) - V_0(t)$$

where S(o) is the initial storage at time t=o, S(t) is the storage at time t, $V_o(t)$ and $V_I(t)$ are volumes of outflow and inflow at time t. Each of the terms of this lumped equation is the result of a number of other terms. These can be sub-divided and even eliminated from the equation depending upon the temporal and spatial scale of the study. The continuity equation is one of the governing equations of almost all hydrologic problems. For a watershed, earlier equation may be written as

$$P + Q_{SI} + Q_{GI} - E - Q_{SO} - Q_{GO} - \Delta S - \varepsilon = 0$$

where P is precipitation, Q_{SI} is surface inflow, Q_{GI} is ground water inflow, E is evaporation from the watershed, Q_{SO} is surface water outflow, Q_{GO} is ground water outflow, ΔS is change in the storage of water in the watershed, and ε is a discrepancy term. For large watersheds, Q_{GI} and Q_{GO} are usually negligible. The discrepancy term is included in equation because the sum of all other terms may not be zero due to measurement errors and/or simplifying assumptions. However, a small value of discrepancy term does not necessarily means that all other terms have been correctly measured/estimated.

Depending on the specific problem, the terms of earlier equation may be further subdivided. For example, when applying the hydrologic equation for short time intervals, the change in the total water storage (ΔS) may be subdivided into several parts: changes of moisture storage in the soil (ΔM), in aquifers (ΔG), in lakes and reservoirs (ΔL), in river channels (ΔS_c), in glaciers (ΔS_G), and in snow cover (ΔS_S). Thus, ΔS can be expressed as:

$$\Delta S = \Delta M + \Delta G + \Delta L + \Delta S_C + \Delta S_G + \Delta S_S$$

The hydrologic equation may be applied for any time interval; the computation of the mean annual water balance for a basin being the simplest, since it is possible to disregard changes in water storages in the basin (ΔS), which are difficult to measure and compute. In general, the shorter the time interval, the more stringent are the requirements for measurement or computation of the components and the more subdivided are the terms of earlier equation. This results in a complex equation which is difficult to close with acceptable errors.

The hydrologic equation may be applied for areas of any size, but the complexity of computation greatly depends on the extent of the area under study. The smaller is the area, the more complicated is its water balance because it is difficult to estimate components of the equation. Finally, the components of the hydrologic equation may be expressed in terms of the mean depth of water (mm), or as a volume of water (m³), or in the form of flow rates (m³/s or mm/s).

Global Water Balance

According to estimates, the annual average depth of precipitation on the land surface is about $108*10^3$ km³. Out of this, about $61*10^3$ km³ is returned to the atmosphere as evapotranspiration and the runoff from land to oceans is $47*103$ km³. As far as the water balance of oceans is concerned, the depth of precipitation over them is about $410*10^3$ km³, $47*103$ km³ of water is received as runoff from the land, and $457*10^3$ km³ is lost as evaporation. If we consider the water balance of atmosphere, $457*10^3$ km³ of water is received as evaporation from oceans and $61*10^3$ km³ from land. The precipitation over oceans is $410*10^3$ km³ and it is $108*10^3$ km³ over land.

Modeling Philosophy

A model is a representation of reality in simple form based on hypotheses and equations:

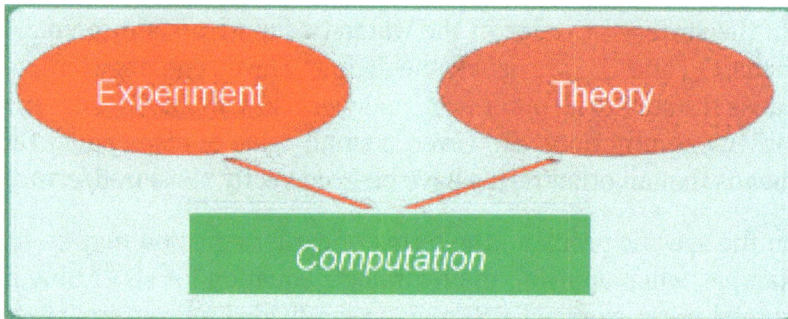

There are two types of models:

- Conceptual
- Mathematical

Conceptual Model

A conceptual model is a representation of a system, made of the composition of concepts which are used to help people know, understand, or simulate a subject the model represents. Some models are physical objects; for example, a toy model which may be assembled, and may be made to work like the object it represents.

The term *conceptual model* may be used to refer to models which are formed after a conceptualization or generalization process. Conceptual models are often abstractions of things in the real world whether physical or social. Semantics studies are relevant to various stages of concept formation and use as Semantics is basically about concepts, the meaning that thinking beings give to various elements of their experience.

Models of Concepts and Models that are Conceptual

The term *conceptual model* is normal. It could mean "a model of concept" or it could mean "a model that is conceptual." A distinction can be made between *what models are* and *what models are models of.* With the exception of iconic models, such as a scale model of Winchester Cathedral, most models are concepts. But they are, mostly, intended to be models of real world states of affairs. The value of a model is usually directly proportional to how well it corresponds to a past, present, future, actual or potential state of affairs. A model of a concept is quite different because in order to be a good model it need not have this real world correspondence. In artificial intelligence conceptual models and conceptual graphs are used for building expert systems and knowledge-based systems; here the analysts are concerned to represent expert opinion on what is true not their own ideas on what is true.

Type and Scope of Conceptual Models

Conceptual models (models that are conceptual) range in type from the more concrete, such as the mental image of a familiar physical object, to the formal generality and abstractness of mathematical models which do not appear to the mind as an image. Conceptual models also range in terms of the scope of the subject matter that they are taken to represent. A model may, for instance, represent a single thing (e.g. the *Statue of Liberty*), whole classes of things (e.g. *the electron*), and even very vast domains of subject matter such as *the physical universe*. The variety and scope of conceptual models is due to then variety of purposes had by the people using them.

Conceptual modeling is the activity of formally describing some aspects of the physical and social world around us for the purposes of understanding and communication.

Fundamental Objectives

A conceptual model's primary objective is to convey the fundamental principles and basic functionality of the system which it represents. Also, a conceptual model must be developed in such a way as to provide an easily understood system interpretation for the models users. A conceptual model, when implemented properly, should satisfy four fundamental objectives.

1. Enhance an individual's understanding of the representative system

2. Facilitate efficient conveyance of system details between stakeholders

3. Provide a point of reference for system designers to extract system specifications

4. Document the system for future reference and provide a means for collaboration

The conceptual model plays an important role in the overall system development life cycle. Figure below, depicts the role of the conceptual model in a typical system development scheme. It is clear that if the conceptual model is not fully developed, the execution of fundamental system properties may not be implemented properly, giving way to future problems or system shortfalls. These failures do occur in the industry and have been linked to; lack of user input, incomplete or unclear requirements, and changing requirements. Those weak links in the system design and development process can be traced to improper execution of the fundamental objectives of conceptual modeling. The importance of conceptual modeling is evident when such systemic failures are mitigated by thorough system development and adherence to proven development objectives/techniques.

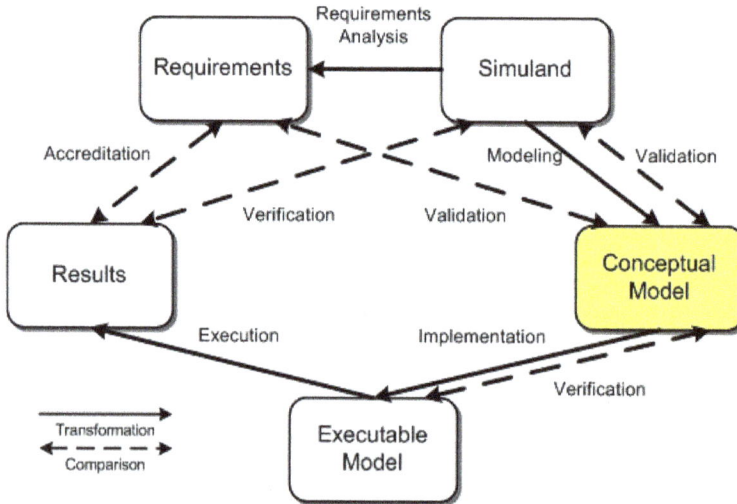

Modeling Techniques

As systems have become increasingly complex, the role of conceptual modeling has dramatically expanded. With that expanded presence, the effectiveness of conceptual modeling at capturing the fundamentals of a system is being realized. Building on that realization, numerous conceptual modeling techniques have been created. These techniques can be applied across multiple disciplines to increase the users understanding of the system to be modeled. A few techniques are briefly described in the following text, however, many more exist or are being developed. Some commonly used conceptual modeling techniques and methods include: workflow modeling, workforce modeling, rapid application development, object-role modeling, and the Unified Modeling Language (UML).

Data Flow Modeling

Data flow modeling (DFM) is a basic conceptual modeling technique that graphically represents elements of a system. DFM is a fairly simple technique, however, like many conceptual modeling techniques, it is possible to construct higher and lower level representative diagrams. The data flow diagram usually does not convey complex system details such as parallel development considerations or timing information, but rather works to bring the major system functions into context. Data flow modeling is a central technique used in systems development that utilizes the structured systems analysis and design method (SSADM).

Entity Relationship Modeling

Entity-relationship modeling (ERM) is a conceptual modeling technique used primarily for software system representation. Entity-relationship diagrams, which are a product of executing the ERM technique, are normally used to represent database models and information systems. The main components of the diagram are the entities and relationships. The entities can represent independent functions, objects, or events. The relationships are responsible for relating the entities to one another. To form a system process, the relationships are combined with the entities and any attributes needed to further describe the process. Multiple diagramming conventions exist for this technique; IDEF1X, Bachman, and EXPRESS, to name a few. These conventions are just different ways of viewing and organizing the data to represent different system aspects.

Event-driven Process Chain

The event-driven process chain (EPC) is a conceptual modeling technique which is mainly used to systematically improve business process flows. Like most conceptual modeling techniques, the event driven process chain consists of entities/elements and functions that allow relationships to be developed and processed. More specifically, the EPC is made up of events which define what state a process is in or the rules by which it operates. In order to progress through events, a function/ active event must be executed. Depending on the process flow, the function has the ability to transform event states or link to other event driven process chains. Other elements exist within an EPC, all of which work together to define how and by what rules the system operates. The EPC technique can be applied to business practices such as resource planning, process improvement, and logistics.

Joint Application Development

The dynamic systems development method (DSDM) uses a specific process called JEFFF to conceptually model a systems life cycle. JEFFF is intended to focus more on the higher level development planning that precedes a projects initialization. The JAD process calls for a series of workshops in which the participants work to identify, define, and generally map a successful project from conception to completion. This method has been found to not work well for large scale applications, however smaller applications usually report some net gain in efficiency.

Place/Transition Net

Also known as Petri nets, this conceptual modeling technique allows a system to be constructed with elements that can be described by direct mathematical means. The petri net, because of its nondeterministic execution properties and well defined mathematical theory, is a useful technique for modeling concurrent system behavior, i.e. simultaneous process executions.

State Transition Modeling

State transition modeling makes use of state transition diagrams to describe system behavior. These state transition diagrams use distinct states to define system behavior and changes. Most current modeling tools contain some kind of ability to represent state transition modeling. The use of state transition models can be most easily recognized as logic state diagrams and directed graphs for finite-state machines.

Technique Evaluation and Selection

Because the conceptual modeling method can sometimes be purposefully vague to account for a broad area of use, the actual application of concept modeling can become difficult. To alleviate this issue, and shed some light on what to consider when selecting an appropriate conceptual modeling technique, the framework proposed by Gemino and Wand will be discussed in the following text. However, before evaluating the effectiveness of a conceptual modeling technique for a particular application, an important concept must be understood; Comparing conceptual models by way of specifically focusing on their graphical or top level representations is shortsighted. Gemino and Wand make a good point when arguing that the emphasis should be placed on a conceptual modeling language when choosing an appropriate technique. In general, a conceptual model is developed using some form of conceptual modeling technique. That technique will utilize a conceptual modeling language that determines the rules for how the model is arrived at. Understanding the capabilities of the specific language used is inherent to properly evaluating a conceptual modeling technique, as the language reflects the techniques descriptive ability. Also, the conceptual modeling language will directly influence the depth at which the system is capable of being represented, whether it be complex or simple.

Considering Affecting Factors

Building on some of their earlier work, Gemino and Wand acknowledge some main points to consider when studying the affecting factors: the content that the conceptual model must represent, the method in which the model will be presented, the characteristics of the models users, and the conceptual model languages specific task. The conceptual models content should be considered in order to select a technique that would allow relevant information to be presented. The presentation method for selection purposes would focus on the techniques ability to represent the model at the intended level of depth and detail. The characteristics of the models users or participants is an important aspect to consider. A participant's background and experience should coincide with the conceptual models complexity, else misrepresentation of the system or misunderstanding of key system concepts could lead to problems in that systems realization. The conceptual model language task will further allow an appropriate technique to be chosen. The difference between creating a system conceptual model to convey system functionality and creating a system conceptual model to interpret that functionality could involve to completely different types of conceptual modeling languages.

Considering Affected Variables

Gemino and Wand go on to expand the affected variable content of their proposed framework by considering the focus of observation and the criterion for comparison. The focus of observation considers whether the conceptual modeling technique will create a "new product", or whether the technique will only bring about a more intimate understanding of the system being modeled. The criterion for comparison would weigh the ability of the conceptual modeling technique to be efficient or effective. A conceptual modeling technique that allows for development of a system model which takes all system variables into account at a high level may make the process of understanding the system functionality more efficient, but the technique lacks the necessary information to explain the internal processes, rendering the model less effective.

When deciding which conceptual technique to use, the recommendations of Gemino and Wand can be applied in order to properly evaluate the scope of the conceptual model in question. Understanding the conceptual models scope will lead to a more informed selection of a technique that properly addresses that particular model. When deciding between modeling techniques, answering the following questions would allow one to address some important conceptual modeling considerations.

1. What content will the conceptual model represent?

2. How will the conceptual model be presented?

3. Who will be using or participating in the conceptual model?

4. How will the conceptual model describe the system?

5. What is the conceptual models focus of observation?

6. Will the conceptual model be efficient or effective in describing the system?

Another function of the simulation conceptual model is to provide a rational and factual basis for assessment of simulation application appropriateness.

Models in Philosophy and Science

Mental Model

In cognitive psychology and philosophy of mind, a mental model is a representation of something in the mind, but a mental model may also refer to a nonphysical external model of the mind itself.

Metaphysical Models

A metaphysical model is a type of conceptual model which is distinguished from other conceptual models by its proposed scope; a metaphysical model intends to represent reality in the broadest possible way. This is to say that it explains the answers to fundamental questions such as whether matter and mind are one or two substances; or whether or not humans have free will.

Epistemological Models

An epistemological model is a type of conceptual model whose proposed scope is the known and the knowable, and the believed and the believable.

Logical Models

In logic, a model is a type of interpretation under which a particular statement is true. Logical models can be broadly divided into ones which only attempt to represent concepts, such as mathematical models; and ones which attempt to represent physical objects, and factual relationships, among which are scientific models.

Model theory is the study of (classes of) mathematical structures such as groups, fields, graphs, or even universes of set theory, using tools from mathematical logic. A system that gives meaning to the sentences of a formal language is called a model for the language. If a model for a language moreover satisfies a particular sentence or theory (set of sentences), it is called a model of the sentence or theory. Model theory has close ties to algebra and universal algebra.

Mathematical Models

Mathematical models can take many forms, including but not limited to dynamical systems, statistical models, differential equations, or game theoretic models. These and other types of models can overlap, with a given model involving a variety of abstract structures.

A more comprehensive type of mathematical model uses a linguistic version of category theory to model a given situation. Akin to entity-relationship models, custom categories or sketches can be directly translated into database schemas. The difference is that logic is replaced by category theory, which brings powerful theorems to bear on the subject of modeling, especially useful for translating between disparate models (as functors between categories).

Scientific Models

A scientific model is a simplified abstract view of a complex reality. A scientific model represents empirical objects, phenomena, and physical processes in a logical way. Attempts to formalize the principles of the empirical sciences use an interpretation to model reality, in the same way logicians axiomatize the principles of logic. The aim of these attempts is to construct a formal system for which reality is the only interpretation. The world is an interpretation (or model) of these sciences, only insofar as these sciences are true.

Statistical Models

A statistical model is a probability distribution function proposed as generating data. In a parametric model, the probability distribution function has variable parameters, such as the mean and variance in a normal distribution, or the coefficients for the various exponents of the independent variable in linear regression. A nonparametric model has a distribution function without parameters, such as in bootstrapping, and is only loosely confined by assumptions. Model selection is a statistical method for selecting a distribution function within a class of them, e.g., in linear regression where the dependent variable is a polynomial of the independent variable with parametric coefficients, model selection is selecting the highest exponent, and may be done with nonparametric means, such as with cross validation.

In statistics there can be models of mental events as well as models of physical events. For example, a statistical model of customer behavior is a model that is conceptual (because behavior is physical), but a statistical model of customer satisfaction is a model of a concept (because satisfaction is a mental not a physical event).

Social and Political Models

Economic Models

In economics, a model is a theoretical construct that represents economic processes by a set of variables and a set of logical and/or quantitative relationships between them. The economic model is a simplified framework designed to illustrate complex processes, often but not always using mathematical techniques. Frequently, economic models use structural parameters. Structural parameters are underlying parameters in a model or class of models. A model may have various parameters and those parameters may change to create various properties.

Models in Systems Architecture

A system model is the conceptual model that describes and represents the structure, behavior, and more views of a system. A system model can represent multiple views of a system by using two different approaches. The first one is the non-architectural approach and the second one is the architectural approach. The non-architectural approach respectively picks a model for each view. The architectural approach, also known as system architecture, instead of picking many heterogeneous and unrelated models, will use only one integrated architectural model.

Business Process Modelling

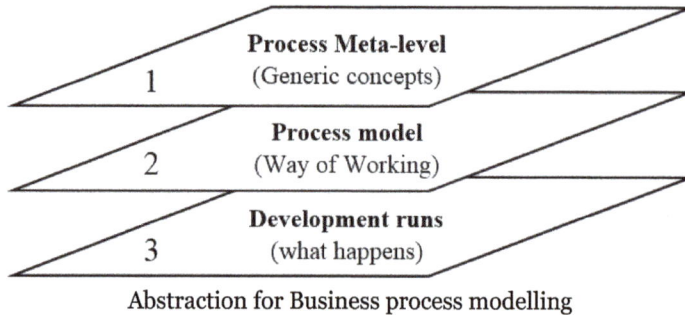

Abstraction for Business process modelling

In business process modelling the enterprise process model is often referred to as the *business process model*. Process models are core concepts in the discipline of process engineering. Process models are:

- Processes of the same nature that are classified together into a model.

- A description of a process at the type level.

- Since the process model is at the type level, a process is an instantiation of it.

The same process model is used repeatedly for the development of many applications and thus, has many instantiations.

One possible use of a process model is to prescribe how things must/should/could be done in contrast to the process itself which is really what happens. A process model is roughly an anticipation of what the process will look like. What the process shall be will be determined during actual system development.

Models in Information System Design

Conceptual Models of Human Activity Systems

Conceptual models of human activity systems are used in soft systems methodology (SSM), which is a method of systems analysis concerned with the structuring of problems in management. These models are models of concepts; the authors specifically state that they are not intended to represent a state of affairs in the physical world. They are also used in information requirements analysis (IRA) which is a variant of SSM developed for information system design and software engineering.

Logico-linguistic Models

Logico-linguistic modeling is another variant of SSM that uses conceptual models. However, this method combines models of concepts with models of putative real world objects and events. It is a graphical representation of modal logic in which modal operators are used to distinguish statement about concepts from statements about real world objects and events.

Data Models

Entity-relationship Model

In software engineering, an entity-relationship model (ERM) is an abstract and conceptual representation of data. Entity-relationship modeling is a database modeling method, used to produce a type of conceptual schema or semantic data model of a system, often a relational database, and its requirements in a top-down fashion. Diagrams created by this process are called entity-relationship diagrams, ER diagrams, or ERDs.

Entity-relationship models have had wide application in the building of information systems intended to support activities involving objects and events in the real world. In these cases they are models that are conceptual. However, this modeling method can be used to build computer games or a family tree of the Greek Gods, in these cases it would be used to model concepts.

Domain Model

A domain model is a type of conceptual model used to depict the structural elements and their conceptual constraints within a domain of interest (sometimes called the problem domain). A domain model includes the various entities, their attributes and relationships, plus the constraints governing the conceptual integrity of the structural model elements comprising that problem domain. A domain model may also include a number of conceptual views, where each view is pertinent to a particular subject area of the domain or to a particular subset of the domain model which is of interest to a stakeholder of the domain model.

Like entity-relationship models, domain models can be used to model concepts or to model real world objects and events.

Mathematical Model

A mathematical model is a description of a system using mathematical concepts and language. The process of developing a mathematical model is termed mathematical modeling. Mathematical models are used in the natural sciences (such as physics, biology, earth science, meteorology) and engineering disciplines (such as computer science, artificial intelligence), as well as in the social sciences (such as economics, psychology, sociology, political science). Physicists, engineers, statisticians, operations research analysts, and economists use mathematical models most extensively. A model may help to explain a system and to study the effects of different components, and to make predictions about behaviour.

Elements of a Mathematical Model

Mathematical models can take many forms, including dynamical systems, statistical models, differential equations, or game theoretic models. These and other types of models can overlap, with a given model involving a variety of abstract structures. In general, mathematical models may include logical models. In many cases, the quality of a scientific field depends on how well the mathematical models developed on the theoretical side agree with results of repeatable experiments. Lack of agreement between theoretical mathematical models and experimental measurements often leads to important advances as better theories are developed.

In the physical sciences, the traditional mathematical model contains four major elements. These are

1. Governing equations

2. Defining equations

3. Constitutive equations

4. Constraints

Classifications

Mathematical models are usually composed of relationships and *variables*. Relationships can be described by *operators*, such as algebraic operators, functions, differential operators, etc. Variables are abstractions of system parameters of interest, that can be quantified. Several classification criteria can be used for mathematical models according to their structure:

- Linear vs. nonlinear: If all the operators in a mathematical model exhibit linearity, the resulting mathematical model is defined as linear. A model is considered to be nonlinear otherwise. The definition of linearity and nonlinearity is dependent on context, and linear models may have nonlinear expressions in them. For example, in a statistical linear model, it is assumed that a relationship is linear in the parameters, but it may be nonlinear in the predictor variables. Similarly, a differential equation is said to be linear if it can be written with linear differential operators, but it can still have nonlinear expressions in it. In a mathematical programming model, if the objective functions and constraints are represented entirely by linear equations, then the model is regarded as a linear model. If one or more of the objective functions or constraints are represented with a nonlinear equation, then the model is known as a nonlinear model. Nonlinearity, even in fairly simple systems, is often associated with phenomena such as chaos and irreversibility. Although there are exceptions, nonlinear systems and models tend to be more difficult to study than linear ones. A common approach to nonlinear problems is linearization, but this can be problematic if one is trying to study aspects such as irreversibility, which are strongly tied to nonlinearity.

- Static vs. dynamic: A *dynamic* model accounts for time-dependent changes in the state of the system, while a *static* (or steady-state) model calculates the system in equilibrium, and thus is time-invariant. Dynamic models typically are represented by differential equations or difference equations.

- Explicit vs. implicit: If all of the input parameters of the overall model are known, and the output parameters can be calculated by a finite series of computations, the model is said to be *explicit*. But sometimes it is the *output* parameters which are known, and the corresponding inputs must be solved for by an iterative procedure, such as Newton's method (if the model is linear) or Broyden's method (if non-linear). In such a case the model is said to be *implicit*. For example, a jet engine's physical properties such as turbine and nozzle throat areas can be explicitly calculated given a design thermodynamic cycle (air and fuel flow rates, pressures, and temperatures) at a specific flight condition and power setting, but the engine's operating cycles at other flight conditions and power settings cannot be explicitly calculated from the constant physical properties.

- Discrete vs. continuous: A discrete model treats objects as discrete, such as the particles in a molecular model or the states in a statistical model; while a continuous model represents the objects in a continuous manner, such as the velocity field of fluid in pipe flows, temperatures and stresses in a solid, and electric field that applies continuously over the entire model due to a point charge.

- Deterministic vs. probabilistic (stochastic): A deterministic model is one in which every set of variable states is uniquely determined by parameters in the model and by sets of previous states of these variables; therefore, a deterministic model always performs the same way for a given set of initial conditions. Conversely, in a stochastic model—usually called a "statistical model"—randomness is present, and variable states are not described by unique values, but rather by probability distributions.

- Deductive, inductive, or floating: A deductive model is a logical structure based on a theory. An inductive model arises from empirical findings and generalization from them. The floating model rests on neither theory nor observation, but is merely the invocation of expected structure. Application of mathematics in social sciences outside of economics has been criticized for unfounded models. Application of catastrophe theory in science has been characterized as a floating model.

Significance in the Natural Sciences

Mathematical models are of great importance in the natural sciences, particularly in physics. Physical theories are almost invariably expressed using mathematical models.

Throughout history, more and more accurate mathematical models have been developed. Newton's laws accurately describe many everyday phenomena, but at certain limits relativity theory and quantum mechanics must be used; even these do not apply to all situations and need further refinement. It is possible to obtain the less accurate models in appropriate limits, for example relativistic mechanics reduces to Newtonian mechanics at speeds much less than the speed of light. Quantum mechanics reduces to classical physics when the quantum numbers are high. For example, the de Broglie wavelength of a tennis ball is insignificantly small, so classical physics is a good approximation to use in this case.

It is common to use idealized models in physics to simplify things. Massless ropes, point particles, ideal gases and the particle in a box are among the many simplified models used in physics. The laws of physics are represented with simple equations such as Newton's laws, Maxwell's equations and the Schrödinger equation. These laws are such as a basis for making mathematical models of real situations. Many real situations are very complex and thus modeled approximate on a computer, a model that is computationally feasible to compute is made from the basic laws or from approximate models made from the basic laws. For example, molecules can be modeled by molecular orbital models that are approximate solutions to the Schrödinger equation. In engineering, physics models are often made by mathematical methods such as finite element analysis.

Different mathematical models use different geometries that are not necessarily accurate descriptions of the geometry of the universe. Euclidean geometry is much used in classical physics, while special relativity and general relativity are examples of theories that use geometries which are not Euclidean.

Some Applications

Since prehistorical times simple models such as maps and diagrams have been used.

Often when engineers analyze a system to be controlled or optimized, they use a mathematical model. In analysis, engineers can build a descriptive model of the system as a hypothesis of how the system could work, or try to estimate how an unforeseeable event could affect the system. Similarly, in control of a system, engineers can try out different control approaches in simulations.

A mathematical model usually describes a system by a set of variables and a set of equations that establish relationships between the variables. Variables may be of many types; real or integer numbers, boolean values or strings, for example. The variables represent some properties of the system, for example, measured system outputs often in the form of signals, timing data, counters, and event occurrence (yes/no). The actual model is the set of functions that describe the relations between the different variables.

Building Blocks

In business and engineering, mathematical models may be used to maximize a certain output. The system under consideration will require certain inputs. The system relating inputs to outputs depends on other variables too: decision variables, state variables, exogenous variables, and random variables.

Decision variables are sometimes known as independent variables. Exogenous variables are sometimes known as parameters or constants. The variables are not independent of each other as the state variables are dependent on the decision, input, random, and exogenous variables. Furthermore, the output variables are dependent on the state of the system (represented by the state variables).

Objectives and constraints of the system and its users can be represented as functions of the output variables or state variables. The objective functions will depend on the perspective of the model's user. Depending on the context, an objective function is also known as an *index of performance*, as it is some measure of interest to the user. Although there is no limit to the number of objective functions and constraints a model can have, using or optimizing the model becomes more involved (computationally) as the number increases.

For example, economists often apply linear algebra when using input-output models. Complicated mathematical models that have many variables may be consolidated by use of vectors where one symbol represents several variables.

A Priori Information

To analyse something with a typical "black box approach", only the behavior of the stimulus/response will be accounted for, to infer the (unknown) *box*. The usual representation of this *black box system* is a data flow diagram centered in the box.

Mathematical modeling problems are often classified into black box or white box models, according to how much a priori information on the system is available. A black-box model is a system of which there is no a priori information available. A white-box model (also called glass box or clear box) is a system where all necessary information is available. Practically all systems are somewhere between the black-box and white-box models, so this concept is useful only as an intuitive guide for deciding which approach to take.

Usually it is preferable to use as much a priori information as possible to make the model more accurate. Therefore, the white-box models are usually considered easier, because if you have used the information correctly, then the model will behave correctly. Often the a priori information comes in forms of knowing the type of functions relating different variables. For example, if we make a model of how a medicine works in a human system, we know that usually the amount of medicine in the blood is an exponentially decaying function. But we are still left with several unknown parameters; how rapidly does the medicine amount decay, and what is the initial amount of medicine in blood? This example is therefore not a completely white-box model. These parameters have to be estimated through some means before one can use the model.

In black-box models one tries to estimate both the functional form of relations between variables and the numerical parameters in those functions. Using a priori information we could end up, for example, with a set of functions that probably could describe the system adequately. If there is no a priori information we would try to use functions as general as possible to cover all different models. An often used approach for black-box models are neural networks which usually do not make assumptions about incoming data. Alternatively the NARMAX (Nonlinear AutoRegressive Moving Average model with eXogenous inputs) algorithms which were developed as part of nonlinear system identification can be used to select the model terms, determine the model structure, and estimate the unknown parameters in the presence of correlated and nonlinear noise. The advantage of NARMAX models compared to neural networks is that NARMAX produces models that can be written down and related to the underlying process, whereas neural networks produce an approximation that is opaque.

Subjective Information

Sometimes it is useful to incorporate subjective information into a mathematical model. This can be done based on intuition, experience, or expert opinion, or based on convenience of mathematical form. Bayesian statistics provides a theoretical framework for incorporating such subjectivity into a rigorous analysis: we specify a prior probability distribution (which can be subjective), and then update this distribution based on empirical data.

An example of when such approach would be necessary is a situation in which an experimenter bends a coin slightly and tosses it once, recording whether it comes up heads, and is then given the task of predicting the probability that the next flip comes up heads. After bending the coin, the true probability that the coin will come up heads is unknown; so the experimenter would need to make a decision (perhaps by looking at the shape of the coin) about what prior distribution to use. Incorporation of such subjective information might be important to get an accurate estimate of the probability.

Complexity

In general, model complexity involves a trade-off between simplicity and accuracy of the model. Occam's razor is a principle particularly relevant to modeling, its essential idea being that among models with roughly equal predictive power, the simplest one is the most desirable. While added complexity usually improves the realism of a model, it can make the model difficult to understand and analyze, and can also pose computational problems, including numerical instability. Thomas Kuhn argues that as science progresses, explanations tend to become more complex before a paradigm shift offers radical simplification.

For example, when modeling the flight of an aircraft, we could embed each mechanical part of the aircraft into our model and would thus acquire an almost white-box model of the system. However, the computational cost of adding such a huge amount of detail would effectively inhibit the usage of such a model. Additionally, the uncertainty would increase due to an overly complex system, because each separate part induces some amount of variance into the model. It is therefore usually appropriate to make some approximations to reduce the model to a sensible size. Engineers often can accept some approximations in order to get a more robust and simple model. For example, Newton's classical mechanics is an approximated model of the real world. Still, Newton's model is quite sufficient for most ordinary-life situations, that is, as long as particle speeds are well below the speed of light, and we study macro-particles only.

Training

Any model which is not pure white-box contains some parameters that can be used to fit the model to the system it is intended to describe. If the modeling is done by a neural network or other machine learning, the optimization of parameters is called *training*, while the optimization of model hyperparameters is called *tuning* and often uses cross-validation. In more conventional modeling through explicitly given mathematical functions, parameters are often determined by *curve fitting*.

Model Evaluation

A crucial part of the modeling process is the evaluation of whether or not a given mathematical model describes a system accurately. This question can be difficult to answer as it involves several different types of evaluation.

Fit to Empirical Data

Usually the easiest part of model evaluation is checking whether a model fits experimental measurements or other empirical data. In models with parameters, a common approach to test this fit is to split the data into two disjoint subsets: training data and verification data. The training data are used to estimate the model parameters. An accurate model will closely match the verification data even though these data were not used to set the model's parameters. This practice is referred to as cross-validation in statistics.

Defining a metric to measure distances between observed and predicted data is a useful tool of assessing model fit. In statistics, decision theory, and some economic models, a loss function plays a similar role.

While it is rather straightforward to test the appropriateness of parameters, it can be more difficult to test the validity of the general mathematical form of a model. In general, more mathematical tools have been developed to test the fit of statistical models than models involving differential equations. Tools from non-parametric statistics can sometimes be used to evaluate how well the data fit a known distribution or to come up with a general model that makes only minimal assumptions about the model's mathematical form.

Scope of the Model

Assessing the scope of a model, that is, determining what situations the model is applicable to, can be less straightforward. If the model was constructed based on a set of data, one must determine for which systems or situations the known data is a "typical" set of data.

The question of whether the model describes well the properties of the system between data points is called interpolation, and the same question for events or data points outside the observed data is called extrapolation.

As an example of the typical limitations of the scope of a model, in evaluating Newtonian classical mechanics, we can note that Newton made his measurements without advanced equipment, so he could not measure properties of particles travelling at speeds close to the speed of light. Likewise, he did not measure the movements of molecules and other small particles, but macro particles only. It is then not surprising that his model does not extrapolate well into these domains, even though his model is quite sufficient for ordinary life physics.

Philosophical Considerations

Many types of modeling implicitly involve claims about causality. This is usually (but not always) true of models involving differential equations. As the purpose of modeling is to increase our understanding of the world, the validity of a model rests not only on its fit to empirical observations, but also on its ability to extrapolate to situations or data beyond those originally described in the model. One can think of this as the differentiation between qualitative and quantitative predictions. One can also argue that a model is worthless unless it provides some insight which goes beyond what is already known from direct investigation of the phenomenon being studied.

An example of such criticism is the argument that the mathematical models of optimal foraging theory do not offer insight that goes beyond the common-sense conclusions of evolution and other basic principles of ecology.

Examples

- One of the popular examples in computer science is the mathematical models of various machines, an example is the deterministic finite automaton (DFA) which is defined as an abstract mathematical concept, but due to the deterministic nature of a DFA, it is implementable in hardware and software for solving various specific problems. For example, the following is a DFA M with a binary alphabet, which requires that the input contains an even number of 0s.

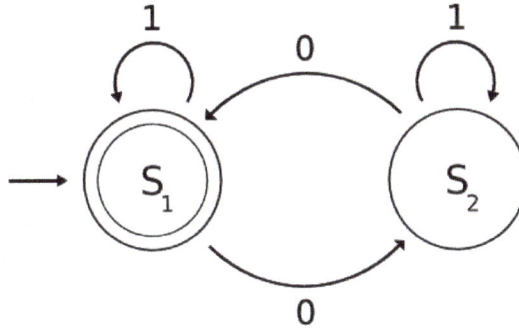

The state diagram for M

$M = (Q, \Sigma, \delta, q_0, F)$ where

- $Q = \{S_1, S_2\}$,

- $\Sigma = \{0, 1\}$,

- $q_0 = S_1$,

- $F = \{S_1\}$, and

- δ is defined by the following state transition table:

	0	1
S_1	S_2	S_1
S_2	S_1	S_2

The state S_1 represents that there has been an even number of 0s in the input so far, while S_2 signifies an odd number. A 1 in the input does not change the state of the automaton. When the input ends, the state will show whether the input contained an even number of 0s or not. If the input did contain an even number of 0s, M will finish in state S_1, an accepting state, so the input string will be accepted.

The language recognized by M is the regular language given by the regular expression 1*(0 (1*) 0 (1*))*, where "*" is the Kleene star, e.g., 1* denotes any non-negative number (possibly zero) of symbols "1".

- Many everyday activities carried out without a thought are uses of mathematical models. A geographical map projection of a region of the earth onto a small, plane surface is a model which can be used for many purposes such as planning travel.

- Another simple activity is predicting the position of a vehicle from its initial position, direction and speed of travel, using the equation that distance traveled is the product of time and speed. This is known as dead reckoning when used more formally. Mathematical modeling in this way does not necessarily require formal mathematics; animals have been shown to use dead reckoning.

- *Population Growth.* A simple (though approximate) model of population growth is the Malthusian growth model. A slightly more realistic and largely used population growth model is the logistic function, and its extensions.

- *Model of a particle in a potential-field.* In this model we consider a particle as being a point of mass which describes a trajectory in space which is modeled by a function giving its coordinates in space as a function of time. The potential field is given by a function $V: \mathbb{R}^3 \to \mathbb{R}$ and the trajectory, that is a function $\mathbf{r}: \mathbb{R} \to \mathbb{R}^3$, is the solution of the differential equation:

$$-\frac{\mathrm{d}^2 \mathbf{r}(t)}{\mathrm{d}t^2} m = \frac{\partial V[\mathbf{r}(t)]}{\partial x} \hat{\mathbf{x}} + \frac{\partial V[\mathbf{r}(t)]}{\partial y} \hat{\mathbf{y}} + \frac{\partial V[\mathbf{r}(t)]}{\partial z} \hat{\mathbf{z}},$$

that can be written also as:

$$m \frac{\mathrm{d}^2 \mathbf{r}(t)}{\mathrm{d}t^2} = -\nabla V[\mathbf{r}(t)].$$

Note this model assumes the particle is a point mass, which is certainly known to be false in many cases in which we use this model; for example, as a model of planetary motion.

- *Model of rational behavior for a consumer.* In this model we assume a consumer faces a choice of n commodities labeled 1,2,...,n each with a market price p_1, p_2,..., p_n. The consumer is assumed to have an ordinal utility function U (ordinal in the sense that only the sign of the differences between two utilities, and not the level of each utility, is meaningful), depending on the amounts of commodities x_1, x_2,..., x_n consumed. The model further assumes that the consumer has a budget M which is used to purchase a vector x_1, x_2,..., x_n in such a way as to maximize $U(x_1, x_2,..., x_n)$. The problem of rational behavior in this model then becomes an optimization problem, that is:

$$\max U(x_1, x_2, \ldots, x_n)$$

subject to:

$$\sum_{i=1}^{n} p_i x_i \leq M.$$

$$x_i \geq 0 \quad \forall i \in \{1, 2, \ldots, n\}$$

This model has been used in a wide variety of economic contexts, such as in general equilibrium theory to show existence and Pareto efficiency of economic equilibria.

- *Neighbour-sensing model* explains the mushroom formation from the initially chaotic fungal network.

- *Computer science*: models in Computer Networks, data models, surface model,...

- *Mechanics*: movement of rocket model,...

Overall Measures of System Performance

- Mean – Average or expected value

- Variance – Average of squared deviations from the mean value

- Reliability – Probability (satisfactory state)

- Resilience – Probability (satisfactory state following unsatisfactory state)

- Robustness – Adaptability to other than design input conditions

- Vulnerability – Expected magnitude or extent of failurewhen unsatisfactory state occurs

- Consistency- Reliability or uniformity of successive results or events

Classification of Watershed Models

- Based on nature of the algorithms

  ◇ Empirical

  ◇ Conceptual

  ◇ Physically based

- Based on nature of input and uncertainty

  ◇ Deterministic

  ◇ Stochastic

- Based on nature of spatial representation

    ◇ Lumped

    ◇ Distributed

    ◇ Black-box

- Based on type of storm event

    ◇ Single event

    ◇ Continuous event

It Can also be Classified as:

- Physical models

    ◇ Hydrologic models of watersheds

    ◇ Scaled models of ships

- Conceptual

    ◇ Differential equations,

    ◇ Optimization

- Simulation models

Descriptive

That depicts or describes how things actually work, and answers the question, "What is this?"

Prescriptive

Suggest what ought to be done (how things should work) according to an assumption or standard.

Deterministic

Here, every set of variable states is uniquely determined by parameters in the model and by sets of previous states of these variables. Therefore, deterministic models perform the same way for a given set of initial conditions.

Probabilistic (Stochastic)

In a stochastic model, randomness is present, and variable states are not described by unique values, but rather by probability distributions.

Static

A static model does not account for the element of time, while a dynamic model does.

Dynamic

Dynamic models typically are represented with difference equations or differential equations.

Discrete

A discrete model does not take into account the function of time and usually uses time-advance methods, while a Continuous model does.

Deductive, Inductive, or Floating

A deductive model is a logical structure based on a theory. An inductive model arises from empirical findings and generalization from them. The floating model rests on neither theory nor observation, but is merely the invocation of expected structure.

Single Event Model

Single event model are designed to simulate individual storm events and have no capabilities for replenishing soil infiltration capacity and other watershed abstraction.

Continuous

Continuous models typically are represented with f(t) and the changes are reflected over continuous time intervals.

Black Box Models

These models describe mathematically the relation between rainfall and surface runoff without describing the physical process by which they are related. e.g. Unit Hydrograph approach.

Lumped Models

These models occupy an intermediate position between the distributed models and Black Box Models. e.g. Stanford Watershed Model.

Distributed Models

These models are based on complex physical theory, i.e. based on the solution of unsteady flow equations.

Input Variables

Space-time fields of precipitation, temperature, etc.

Parameters

- Size
- Shape
- Physiography
- Climate
- Hydrogeology
- Socioeconomics
- Drainage
- Land use
- Vegetation
- Geology and Soils
- Hydrology

State Variables

Space-time fields of soil moisture, etc.

Equations Variables

- Independent variables
  - ◇ space   x
  - ◇ time   t
- Dependent variables
  - ◇ discharge Q
  - ◇ water level h
- All other variables are function of the independent or dependent variables.

Goals and Objectives

Both goals and objectives are very important to accomplish a project. Goals without objectives can never be accomplished while objectives without goals will never take you to where you want to be.

Goals	Objectives
Vague, less structured	Very concrete, specific and measurable
High level statements that provide overall context of what the project is trying to accomplish	Attainable, realistic and low level statements that describe what the project will deliver.

Hydrologic Data

Hydrological observations are the scientific ways for collection of water related data at a specific location. There are many ways in which the hydrologic data can be collected. The major techniques are described below.

Direct Measurement

This is the most common way to measure hydrometeorological variables, such as precipitation and streamflow. A gauging site is established and is equipped with the devices that can measure the variable(s) of interest. In case of manual observations, an observer visits the site, measures the values of the concerned variables, and records or transmits them to the controlling office for processing and storage. On the other hand, at an automated hydrologic or weather station the seasons can measure a number of hydrometeorological variables and store/transmit the data to the controlling office without any human intervention. The equipment may be programmed to transmit the data at selected time interval or it can be interrogated as per the needs to get the data. With improvement in communication technology, it is possible to get the desired data from the stations widely spread over an area at a central place in real-time.

Remote Sensing

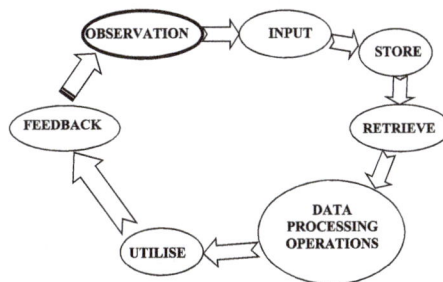

Observation and input, processing and storage, Retrieval and use, Feedback.
An illustration of data processing life cycle activities

In this technique, the data about an object are obtained without coming in physical contact with the object. This technique is now very commonly used to provide spatial data of terrain features. Similarly, weather radars are being increasingly used for measurement of precipitation.

Chemical Tracers

In this approach, some chemicals, known as tracers, are added to the process whose data are to be obtained. Tracers can also be used to determine the flow path of water or a pollutant. The nuclear or isotope techniques are being employed to trace the movement of water molecules in different parts of the hydrological cycle. Nuclear techniques are helpful to assess the rate of sediment deposition in a water body, identify the rainfall recharge and recharge areas of aquifers, study of seawater intrusion in coastal regions, measure seepage and leakage from surface water bodies, analyse surface water and ground water interaction, etc.

Classification of Hydrologic Data

Hydrologic data can be classified in several ways. Most commonly, data are classified in three categories: time-oriented data, space-oriented data, and relation-oriented data.

Hydrologic data can also be classified as time varying or time non-varying data. The time non-varying or static data includes most space-oriented data which do not change (or very-very slowly change) with time, for example catchment topographic map, soil map, etc. Some features, such as river network and land use in a catchment, might gradually change with time and can be considered as semi-static. A brief description of each type of data is presented next.

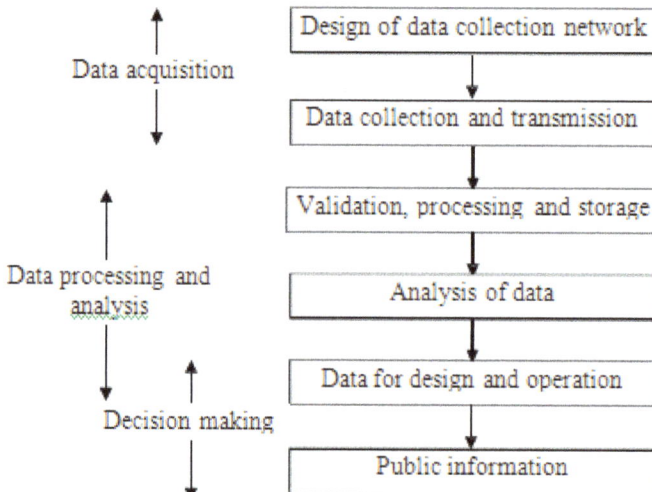

Activities of a hydrological service.

Time-oriented Data

Values of most hydrometeorological variables change with time and such variables are known as time-oriented data. The time-series data include all the measurements which have an observation time associated with them and most water resources data have this property. The variable could be an instantaneous value (e.g. river water level); an accumulated value (e.g., daily rainfall); or an averaged value (e.g., mean daily discharge). The distinction between instantaneous and accumulative values is important when the data are processed. These data can be further classified as meteorological data, hydrological data, and water quality data.

Depending on the frequency of observations, the time-series data can also be classified as:

- Equidistant time-series data which are the measurements made at regular intervals (hourly, daily); the reported values may be instantaneous, accumulated or averaged.

- Cyclic time-series data are the data measured at irregular intervals of time but the time sequence is repeated regularly. For example, the observation of at many places river stage is measured daily at 08:30 and 17:30 hrs.

- Values of non-equidistant data series are observed when some specified event takes place. For example, in a tipping bucket rain gauge, the bucket tips after a certain depth of rain has fallen and the value is recorded.

Space-oriented Data

Space-oriented data comprise of the information related to physical and morphological characteristics of catchments, rivers (cross-sections, profile, bed characteristics, networks), soil maps, lakes and reservoirs data (elevation-area variation), etc. Traditionally, such data are stored in the form of paper maps and manually analyzed. The modern trend is to use a Geographical Information System (GIS) to input, store and analyze such data. Different types of information, such as topographical and land use of an area, are stored in a GIS in different layers of a map which can be overlaid and analyzed.

Relation-oriented Data

Such data comprise of information about mathematical relationships established between two or more variables. A mathematical relationship between two or more variables is established for many purposes, such as data validation, filling-in missing data, etc. The variables themselves may form a time-series but their relationship is of interest here. The relationship may be expressed in mathematical, tabular, or graphical form. The stage-discharge rating, spillway rating table and the calibration ratings of various instruments are typical example of relation-oriented data. More than one equation may be required to characterize the relationship which may change with time.

Hydrological Network Design

Precipitation Networks

The optimum density of a precipitation gauge network depends on the purpose for which the data are to be used. For example, measurements of precipitation for flood forecasting require denser networks as compared to that for rainfall-runoff modeling. WMO (2008) has recommended the following table as minimum network densities for precipitation stations.

Table: Recommended minimum densities of stations (area in km² per station).

Physiographic unit	Precipitation		Evaporation
	Non-recording	Recording	
Coastal	900	9000	50000
Mountains	250	2500	50000
Interior plains	575	5750	5000
Hilly/undulating	575	5750	50000
Small islands	25	250	50000
Urban areas	-	10–20	-
Polar/arid	10000	100000	100000

The optimum network should make it possible to derive required information with desired accuracy. The optimum number of rain gauge stations (N) in a network is given by:

$$N = [C_v / p]^2$$

where C_v = the coefficient of variation of the precipitation values of the existing rain gauge stations, and p = the allowable maximum percentage error in the estimate of basin mean rainfall. A typical value of p is 10 percent. Here, C_v is computed by

$$C_v = 100 * s / P_m$$

In which s is the standard deviation and P_m is the mean rainfall of the existing stations. Obviously, a decrease in the percentage error would mean an increase in the number of gauges required. Mukherjee and Kaur (1987) have proposed a modified form of eq. i.e, $N = [C_v / p]^2$ by including the mean correlation (r) of precipitation over the area

$$N = [C_v / p]^2 (1 - r)$$

WMO recommends that the precipitation (amount and form) should be measured with an accuracy of 3–7% and rainfall intensity with 1 mm/hr at the 95 per cent confidence interval. Snow depth below 20 cm should be accuracy of less than 1 cm and depth above

20 cm should not have more than 10% error. The recommended accuracy for evaporation range 2–5% and for wind speed 0.5 m/sec.

Example: A catchment has 6 rain gauges and the annual rainfall at these has been measured as 750, 540, 465, 493, 421, and 780 mm. Find out the optimum number of rain gauges for the basin if the error of estimation is required to be kept below 10%.

Solution: For the data given, mean = 574.83 mm and standard deviation = 152.59 mm. Thus

$$C_v = 100 * 152.59 / 574.83 = 26.54$$

Hence, using eq.($N = [C_v / p]^2$), the optimum number of rain gauges for the basin (N) is

$$N = (26.54 / 10)^2 = 7.04$$

This means that 7 rain gauge stations are required in the basin and the existing network of 6 rain gauges is slightly inadequate. It needs to be strengthened by adding one new gauge so that the estimate of rainfall depth has stipulated accuracy.

Stream Gauging Networks

A network of stream gauging stations is established in a river basin to provide data required by the hydrologists for planning, development and management of water resources of the basin. The collected data also enables to estimate the principal characteristics of the hydrological regime of the basin.

Every major stream should be gauged near its mouth and its major tributaries should also be gauged as feasible. Naturally, gauging depends on the existing and likely development in the basin. According to WMO, the first gauging station is selected at the most upstream location where the drainage area is about 1300 km². The second station is located at a point in the downstream direction where the drainage area is approximately doubled. The WMO recommendations for a minimum density of stream gauging stations are given in the table.

Table: Recommended minimum densities of stations (area in km² per station).

Physiographic unit	Streamflow	Sediments	Water quality
Coastal	2750	18300	55000
Mountains	1000	6700	20000
Interior plains	1875	12500	37500
Hilly/undulating	1875	12500	47500
Small islands	300	2000	6000
Urban areas	-	-	-
Polar/arid	20000	200000	200000

Stations are also established in the basin at the sites where significant changes in the volume of flow are noticed, for example downstream of the confluence of a major tributary or at the outflow point of a lake etc. In case a suitable location is not available downstream of the confluence, the sites can be located upstream of the confluence, preferably on the tributary. While establishing sites at the downstream of confluence, it should be ensured that no other small stream joins the main river before the station so that correct assessment of the contribution of the tributary to the main river is obtained.

The distance between two consequent stations on the same river may vary from about 50 km to several hundred kilometers, depending on many factors. The drainage area computed from origin up to two consecutive observation sites on a large river should preferably differ by more than ten percent so that the difference in quantities of flow at the two stations is significant. Sometimes stations are set up due to hydrological significance, say, to determine the flow contribution from a typical catchment.

A different approach is recommended for small independent rivers which flow directly into the sea (for example, the rivers in Western Ghats). In such cases, the first hydrological observation station is to be established on a stream that is typical of the region and further stations are added to the network to cover the area and obtain information about the variability. Stream in the area whose flows are low should not be avoided from the network. Absence of stations from low flow streams may lead to wrong assessment of the surface water potential of the area if it has been evaluated just on the basis of the data from the high flow streams. Thus, great care is to be exercised to ensure that all distinct hydrological features are adequately covered by the gauging network.

An existing gauging network needs periodic review. The developments that take place in the basin like construction of new water resources development projects may warrant addition or closure of the sites. Often the rivers are polluted by the discharge of affluents from industries. A need may also arise to establish stations to monitor the quality of water in the river.

Regarding the accuracy desired in measuring river water depth and discharge, WMO recommends that the water depth measurement should have accuracy of about 2%, velocity of flow 2–5%, and discharge about 5%. Suspended sediment concentration should be estimated with accuracy of 10%.

Validation of Hydrologic Data

Errors in Hydrological Observations

Theoretically, the true values of hydrological variables cannot be determined by measurements because errors of measurement cannot be eliminated completely. Errors arise in hydrometric measurements due to several reasons. Most common causes are:

- Faulty equipment, e.g., a current meter with worn-out or damages parts,

- Malfunction of instrument, e.g., slippage of float tape in water level recorder,

- Improper exposure conditions of the instruments e.g., a rain gauge surrounded by high rise buildings,

- Observation errors by the observer, e.g., gauge misread,

- Wrong entry of data in records/computer and

- Error in computation, e.g., mistake while converting current meter rotations to velocity.

Table: A partial list of ISO standards related to Hydrometry.

ISO Number	Details
ISO 748:2007	Hydrometry -- Measurement of liquid flow in open channels using current meters or floats
ISO 772:2011	Hydrometry -- Vocabulary and symbols
ISO 1070:1992	Liquid flow measurement in open channels -- Slope-area method
ISO 1100-2:2010	Hydrometry -- Measurement of liquid flow in open channels -- Part 2: Determination of the stage-discharge relationship
ISO 1438:2008	Hydrometry -- Open channel flow measurement using thin-plate weirs
ISO 2425: 2010	Hydrometry – Measurement of liquid flow in open channels under tidal conditions
ISO 3846:2008	Hydrometry -- Open channel flow measurement using rectangular broad crested weirs
ISO 4359:1983	Liquid flow measurement in open channels -- Rectangular, trapezoidal and U- shaped flumes
ISO 4360:2008	Hydrometry -- Open channel flow measurement using triangular profile weirs
ISO 4362:1999	Hydrometric determinations -- Flow measurement in open channels using structures -- Trapezoidal broad-crested weirs
ISO 4373:2008	Hydrometry -- Water level measuring devices
ISO 4374:1990	Liquid flow measurement in open channels -- Round-nose horizontal broad-crested weirs

It needs to be stressed that no statistical analysis can replace correct observations because spurious and systematic errors cannot be eliminated by such analysis. Only random errors can be characterized by statistical means.

Errors in hydrologic measurements can be classified in three categories: systematic, random, and spurious. Figure gives a graphical depiction of errors. These are discussed in detail in the following image:

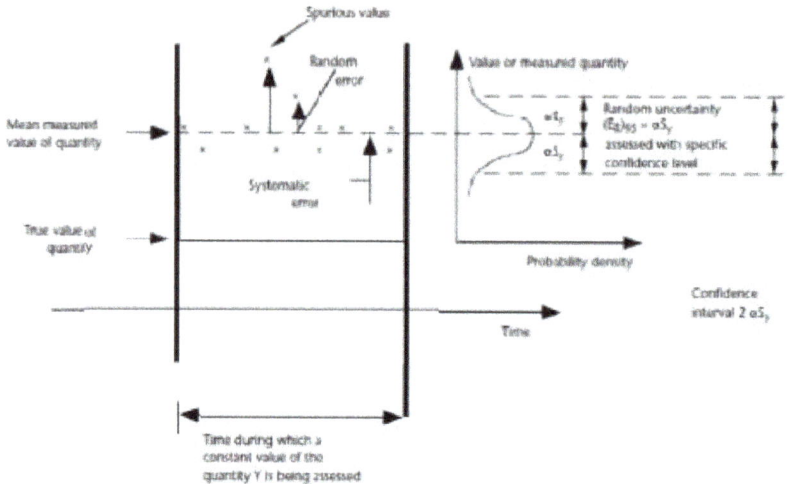

Explanation of measurement errors

Systematic Errors

A *systematic* error or bias is a systematic difference, either positive or negative, between the measured value and the true value. Systematic errors arise mainly due to malfunctioning of instrument. Hence, if the instruments and measurements conditions remain unchanged, such errors cannot be reduced just by increasing the number of measurements. Systematic errors also arise often due to difficult measuring conditions, such as unsteady flow, meandering and bad location of observation stations, and lack of knowledge of observer. Such error should be eliminated by properly adjusting, repairing, or changing the instrument, and by changing the measurement conditions. For example, this can be done by straightening the approach channel of a stream-gauging section. If the systematic error has a known value, this should be accounted for appropriately and error due to this source should be considered zero. Systematic errors are generally more serious and the validation process must be able to detect and correct them.

Regarding precipitation, WMO (1982) listed the following errors for which adjustment needs to be made to get a near accurate estimate of precipitation from a measured precipitation report.

1. error due to the systematic wind field deformation above the gauge orifice

2. error due to the wetting loss on the internal walls of the collector

3. error due to evaporation from the container (generally in hot climates)

4. error due to the wetting loss in the container when it is emptied

5. error due to blowing and drifting snow

6. error due to splashing in and out of water, and

7. random observational and instrumental errors.

The first six errors listed above are systematic and are listed in order of general importance. The net error due to blowing and drifting snow and due to splash in and out of water can be either negative or positive while net systematic errors due to the wind field and other factors are negative. Since for liquid precipitation the errors listed at (5) and (6) above are near zero, the general model for adjusting the data from most gauges takes the form

$$Pk = K (Pg + \Delta P1 + \Delta P2 + \Delta P3)$$

where

Pk = adjusted precipitation amount

K = adjustment factor for the effects of wind field deformation

Pg = the measured amount of precipitation in the gauge

$\Delta P1$ = adjustment for the wetting loss in the internal wells of the collector

$\Delta P2$ = adjustment for wetting loss in the container after emptying

$\Delta P3$ = adjustment for evaporation from the container

The data needed to make the adjustments include wind speed, drop size, precipitation intensity, air temperature, humidity and other characteristic of the gauge site.

Random Errors

Random errors vary in an unpredictable manner, both in magnitude and sign, when repeated measurements of the same variable are made under the same conditions. Random errors cannot be eliminated, but their impacts can be reduced by repeated measurements of the variable. These are equally distributed about the mean or 'true' value. The errors of individual readings may be large or small, e.g., the errors in a staff gauge reading where the water surface is subject to wave action. Usually, they compensate with time or are minimized by taking a sufficient number of measurements. The uncertainty of the arithmetic mean computed from n independent measurements is several times smaller than the uncertainty of a single measurement. The distribution of random errors can usually be assumed to be normal (Gaussian). For certain cases, normal distribution can or should be replaced by other statistical distributions. These errors can be identified by a statistical-outlier test that gives a rejection criterion.

In measuring rainfall, random errors could arise due to spilling of the water when transferring it to the measuring jar, leakage into or out of the receiver, observational error etc. The others random errors which could be due to observer include:

   i.   misreading and transposing digits,

   ii. misrecording because of faulty memory,

   iii. recording the data at the wrong place on the recording sheet,

   iv. making readings at improper interval,

   v.   incorrect dating of the report,

   vi. incorrectly reading or communicating the data to a reporting centre, etc.

It appears, therefore, that computerization and automation may be solution to reduce the error. However, even without human intervention chances of erroneous reading in case of precipitation may be possible because of

   i.   evaporation from gauge,

   ii. overflowing gauge,

   iii. mechanical or electrical mal-functions.

Spurious Errors

These arise due to human mistakes or instrument malfunction or some abnormal external cause. Reported data appear to be clearly in error. Sometimes the errors become obvious, for example, wrong placement of decimal and the data can be easily corrected in such cases. In other cases, the concerned measurements may have to be discarded. For example, an animal may drink water from the evaporation pan and introduce errors in the data. Sometimes, such errors may be readily detected but it may not be easy to correct them.

Sources of Errors

After understanding the types of errors, the next obvious question will be about their sources. It would not be possible to list all likely sources of error because there are different instruments and measuring methods and each of these will have their own sources of error. Some typical sources of error were given by (WMO 2008):

a) Datum or zero error originates from the incorrect determination of the reference point of an instrument, for example, staff-gauge zero level, difference between the staff-gauge zero and the weir-crest levels.

b) Reading (or observation) error results from the incorrect reading of the value indicated by the measuring instrument. This error is normally attributed to neglect or incompetence of the observer. It could also arise, for example, due to bad visibility, waves, or ice at the staff gauge.

c) Interpolation error is due to inexact evaluation of the position of the index with reference to the two adjoining scale marks between which the index is located.

d) Error due to wrong assumption or neglect of one or more variables needed to determine the measured value (for example, assuming a unique stage-discharge relationship during periods of unsteady flow when discharge depends on slope as well as stage).

e) Hysteresis.

f) Insensitivity error arises when the instrument cannot sense the small change in the variable being measured.

g) Non-linearity error is that part of error whereby a change of indication or response departs from proportionality to the corresponding change of the value of the measured quantity over a defined range.

h) Drift error is due to the property of the instrument in which its measurement properties change with time under defined conditions of use, for example, mechanical clockworks drift with time or temperature.

i) Instability error results from the inability of an instrument to maintain certain specified metrological properties constant.

j) Out-of-range error is due to the use of an instrument beyond its effective measuring range, lower than the minimum or higher than the maximum value of the quantity, for which the instrument/installation has been constructed, adjusted, or set (for example, unexpected high water level).

k) Out-of-accuracy class error is due to the improper use of an instrument when the minimum error is more than the tolerance for the measurement.

It may be emphasized here that uncertainty in measurement has a probabilistic character. Therefore, one can define an interval in which the true value of the variables is expected to lie with a certain confidence level. If measurements are independent, then the uncertainty in the measurements can be estimated by taking a minimum of a large number (say > 25 observations) and calculating the standard deviation. A problem in applying statistics to hydrological data is that many hydrological variables are assumed to be independent random variables. For many hydrologic variables, this assumption is not strictly valid. For example, short-term river flows are correlated.

Secondary Errors of Measurement

Many hydrological variables are not directly measured but are estimated form the measured values of several variables. For example, discharge passing through a site may be estimated by the stage at that site or discharge at a weir may be computed as a function of a discharge coefficient, characteristic dimensions and head. If the individual components and their errors are assumed to be statistically independent, the resultant uncertainty (also referred to as overall uncertainty) can be calculated from the uncertainties of the individual variable.

Let a quantity, Q, be a function of three measured quantities, x, y and z, and the uncertainty in Q be denoted by e^Q. Further, let the uncertainties in variables x, y and z be given by e_x, e_y and e_z, respectively. We can estimate the uncertainty e^Q by applying the Gauss error transfer theorem (WMO 2008):

$$(e_q)^2 = \left(\frac{\partial Q}{\partial x}e_x\right)^2 + \left(\frac{\partial Q}{\partial y}e_y\right)^2 + \left(\frac{\partial Q}{\partial z}e_z\right)^2$$

where $\partial Q/\partial x$, $\partial Q/\partial y$ and $\partial Q/\partial z$ are the partial differentials of the function expressing explicitly the relationship of the dependent variable with the independent variables.

Validation of Hydrologic Data

As described above, measured raw data may have errors. Errors may also arise in data entry, during computations and (hopefully very rarely), from the mistaken 'correction' of 'right' data. Reliability of the data determines whether they are suitable in various applications or not. Use of erroneous data may do more harm than good and may introduce an error of unknown order in any decision that has been taken by the use of such data. It is, therefore, necessary that the data are checked for any possible error and these are removed before the data are used in analysis, design, and decision making. Thus, the need for data validation or quality control arises because field measurements are subject to errors. Data validation aims at detecting and removing these likely errors and inconsistencies in the data.

Data validation is the means by which data are checked to ensure that the corrected values are the best possible representation of the true values of the variable. Data validation procedure includes primary and secondary data validation. Validation of hydrologic data must never be considered as a purely statistical or mathematical exercise.

Staff involved in it must have a background in hydrology and must understand the field practices. To understand the source of errors, one must understand the method of measurement or observation in the field and the typical errors of given instruments and techniques. Knowledge of the method of measurement or observation influences our view of why the data are suspect.

Basically, data validation is carried out:

- to correct errors in the observed values where possible,

- to assess the reliability of data even though it may not be possible to correct errors, and

- to identify the source of errors to ensure that these are not repeated in future.

The input variables in an analysis may be directly measured (e.g., rainfall) or they may be derived using a relationship with one or more variables (e.g., discharge that has been obtained from a rating curve). In the latter case the error in the variable (discharge) depends both on field measurements and the error in the relationship. An error may also be introduced if the relationship is no longer valid or the values are extrapolated outside the applicable range. Validation involves different types of comparisons of data and includes the following:

Single series comparison

- between individual observations and pre-set physical limits.

- between sequential observations to detect unacceptable rates of change and deviations from acceptable behaviour (most readily identified graphically) and

- between two measurements of a variable at a single station, e.g., daily rainfall from a daily gauge and an accumulated total from a recording gauge.

Multiple stations/data

- between two or more measurements at nearby stations, e.g. flow at two sites along a river and

- between measurements of different but related variables, e.g., rainfall and river flow.

Levels of Validation

Validation of data is best done soon after observation and at the observation station because secondary or related information to support validation is readily available. However, data validation at observation sites may not be always possible due to logistics and the lack of trained personnel. Validation of hydrological data can be grouped in three major categories: a) primary validation, b) secondary validation, and c) hydrological validation.

Note that none of the procedures of data validation are absolutely objective and there is no guarantee that all the errors will be captured and removed. They are basically tools to screen out suspect data which are to be further examined by other tests and corroborative facts. When it is ascertained that a particular value is incorrect, an alternative value that is likely to be closer to the true value is substituted. Since each hydrological variable has distinct characteristics, it is necessary that specific validation techniques be designed for each variable. Further validation a pure statistical exercise; the properties and behavior of the variable under consideration should always be kept in mind.

Primary Validation

Primary data validation is done to highlight and, if possible, correct those data which are not within the expected range. Primary validation involves comparisons within a single data series or between observations and pre-set limits and/or statistical range of a variable or with the expected behavior of the generating process. Sometimes, information from a few nearby stations may also be pooled. If it is not possible to definitely conclude that the suspected value is erroneous, such value is not changed but is flagged indicating that it is doubtful. All data which have been flagged as suspicious during primary validation are again screened later on the basis of additional information.

Secondary Validation

After primary validation secondary validation of data is taken up to for expected spatial behavior of the variable as inferred from neighboring observation stations. It is assumed that the variable under consideration has spatial correlation within small distances. This assumption must be supported by the underlying behavior of the process under examination. Checks applied at this stage basically examine if the data at the station under consideration is spatially consistent with the data of the surrounding stations. The spatial validation and consistency check is carried out using the data of key stations which are known to be of good quality.

When hydrological variables have a high auto-correlation, such as ground water levels, or the data has high correlation with neighboring stations, the validation and data correction can be carried out with a higher level of confidence. However, processes such as convective rainfall show a great temporal and spatial variability. It is difficult to ascertain the behavior of such processes with the desired degree of confidence. Based on available information and statistical properties, if it is not possible to conclude whether the suspected value is erroneous or not, such value is not changed but is flagged as doubtful. All doubtful data are further validated on the basis of additional information.

Hydrological Validation

Here the basic idea is to correct erroneous data by the use of hydrological knowledge consists of comparing data of correlated variables at nearby stations to identify inconsistencies between the time series or their derived statistics. This test can be applied to a measured variable (water level) or to derived variables (flow, runoff) and is usually done through regression or simulation modeling.

Ideally all the hydrological data should be subjected to hydrological validation. For historical data to which no (or few) checks have been applied, hydrological validation provides an effective check on the quality and reliability of records. Thorough hydrological validation requires a high level of professional expertise and can be time consuming. Required man power and time may not be available always. Therefore, this validation may be applied selectively. Finally, the validation may be able to identify a particular section of record/ data item that is unreliable, but it may not always be possible to correct the values.

Validation of Climatic Data

Climatic data are known to have high spatial correlation depending upon topography. Validation is mainly concerned with spatial comparisons between neighboring stations to identify anomalies in recording at the station. Methods of validation can be classified in two groups.

(i) Single station validation tests for homogeneity,

(ii) Multiple station validation.

Single Series Tests of Homogeneity

Single series testing for homogeneity will normally only be used with long data sets. Series may be inspected graphically for evidence of trend and this may often be a starting point. However, statistical hypothesis testing can be more discriminative in distinguishing between expected variation in a random series and real trend or more abrupt changes in the characteristics of the series with time.

Trend Analysis (Time Series Plot)

A series can be considered homogeneous if there is no significant linear or curvilinear trend in the time series of the climatic element. The presence of trend in the time series can be examined by graphical display and/or by using simple statistical tests. The data are plotted on a linear or semi-logarithmic scale with the climatic variable on the Y-axis and time on the X-axis. The presence or absence of trend may be seen by examination of the time series plot. Mathematically one may fit a linear regression and test the regression coefficients for statistical significance.

Trend generally does not become evident for a number of years and so the tests must be carried out on long data series, often aggregated into annual series. Trend may result from a wide variety of factors including:

- Change of instrumentation

- Change of observation practice or observer

- Local shift in the site of the station

- Growth of vegetation or nearby new buildings affecting exposure of the station

- Effects of new irrigation in the vicinity of the station (affecting humidity, temperature and pan evaporation)

- Effects of the urban heat island with growing urbanisation

- Global climatic change

The presence of trend does not necessarily mean that part of the data are erroneous but that the environmental conditions have changed. Unless there is reason to believe that the trend is due to instrumentation or observation practices or observer, the data should not generally be altered but the existence of trend noted in the station record.

Residual Mass Curve

A residual mass curve represents accumulative departures from the mean. It is a very effective visual method of detecting climatic variability or other inhomogeneities. The residual mass curve can be interpreted as follows:

- an upward curve indicates an above average sequence

- a horizontal curve indicates an about average sequence

- a downward curve indicates a below average sequence

Multiple Stations Validation

The simplest and often the most helpful means to identify anomalies between the data of multiple stations by plotting time series of the data of stations on same or adjacent graphs. This should generally be carried out in the very beginning, before other tests. For climate variables the series will usually be displayed as curves of a variable at two or more stations where measurements have been taken concurrently, for example atmospheric temperature, dry bulb temperature, or sunshine hours. It is important to keep the same scale so that the visual impression is right.

While examining the current data, the plot should include the past time series of sufficient length to ensure that there are no discontinuities between one batch of data received from the station and the next. This will ensure that the data are being entered against the correct station and correct date/time.

For climatic variables, which have strong spatial correlation, such as temperature, the graphs will generally run along closely, with nearly the same variation with the mean separation representing some location factor such as altitude. Abrupt or progressive straying from this pattern will be evident from the comparative plot, which would not necessarily have been perceived at primary validation from the inspection of the single station. An example might be the use of a faulty thermometer, in which there might be an abrupt change in the plot in relation to other stations. An evaporation pan affected by leakage may show a progressive shift as the leak develops. This would permit the data processor to delimit the period over which suspect values should be corrected.

Comparison of series may also help in accepting of values which might have been suspected in primary validation because they fell outside the warning range. Where two or more stations display the same behavior, there is strong evidence to suggest that the values are correct. An example might be the occurrence of an anomalous atmospheric pressure in the vicinity of a tropical cyclone.

Comparison plots provide a simple means of identifying anomalies but not of correcting them. This may be done through correlation or regression analysis, spatial homogeneity testing (nearest neighbor analysis) or double mass analysis.

Residual Series

An alternative method of displaying comparative time series is to plot the differences. This procedure is often applied to river flows along a channel to detect anomalies in the water balance but it may equally be applied to climate variable to detect anomalies and to flag suspect values or sequences.

Let X_1 and X_2 be two variables. The difference series is computed as

$$Y_i = X_{1,i} - X_{2,i}$$

Regression Analysis

Regression analysis is a very commonly used statistical method. In the case of climatic variables where individual or short sequences of anomalous values are present in a spatially conservative series, a simple linear relationship with a neighboring station may well provide a sufficient basis for interpolation.

In a plot of the relationship, the suspect values will generally show up as outliers but such plots provide no indication of the time sequencing of the suspect values. One will not be able to say whether the outliers were scattered or contained in one block. For seasonal data, the relationship should be derived for a period within the same season as the suspect values. (The relationship may change between seasons). The identified suspected values should be removed before deriving the relationship, which may be applied to compute corrected values to replace the suspect ones.

Double Mass Curves

Double mass curve analysis, cumulative plots of variable under consideration at one station and surrounding stations may also be used to show trends or in homogeneities between climate stations but it is usually used with longer, aggregated series. However, in the case of a leaking evaporation pan, described above, the display of a mass curve of daily values for a period commencing some time before leakage commenced, the anomaly will show up as a curvature in the mass curve plot.

This procedure may only be used to correct or replace suspect values where there has been a systematic but constant shift in the variable at the station in question, i.e., where the plot shows two straight lines separated by a break of slope. In this case the correction factor is the ratio of the slope of the adjusted mass curve to the slope of the unadjusted mass curve. Where there has been progressive departure from previous behavior, the slope is not constant as in the case of the leaking evaporation pan, and the method should not be used.

Spatial Homogeneity (Nearest Neighbor Analysis)

This procedure is most commonly used for rainfall but can be used for other variables also. Its advantage for rainfall in comparison to climate is that there are generally more rainfall stations in the vicinity of the target station than there are climate stations. The advantage for some climate variables is that there is less spatial variability and the area over which comparison is permitted may be increased.

Closure

The science of hydrology deals with immense volumes of data of a number of variables. Since data collection is expensive process, it is necessary that the data collection campaigns are carefully planned and executed. Errors may creep in the data due to various causes. Therefore, before the data can be put to use, it needs to be screened and validated to remove these errors. After the data has undergone quality control checks, it is ready for use in planning, design, and operation.

Definitions of Terms Related to Measurement Errors

Definitions of some important terms related to accuracy have been sourced from WMO (2008) and are given below.

Accuracy: The word accuracy is generally used to indicate the closeness or the agreement between an experimentally determined value of a quantity and its true value. It is the extent to which a measurement agrees with the true value. An accurate result closely agrees with the actual value for that quantity. In other words, accuracy tells us how close a measurement is to an accepted standard. Precision describes how well repeated measurements agree with each other. It tells us how close two or more measurements agree. It is worth mentioning here that precision does not necessarily indicate anything about the accuracy of the measurements. An experiment is considered good when it is both precise and accurate.

An experiment is said to have high precision, if it has small random error. It is said to have high accuracy, if it has small systematic error. There may be four possibilities for characterizing the obtained experimental data, as shown in figure (B) precise and accurate, (A) precise and inaccurate, (D) imprecise and accurate, and (C) imprecise and inaccurate. In hydrological observation, the objective is to reduce both systematic and random errors as much as possible. However, for economy of effort, one must try to strike a balance between these two sources of error, giving greater weight to the larger of the two.

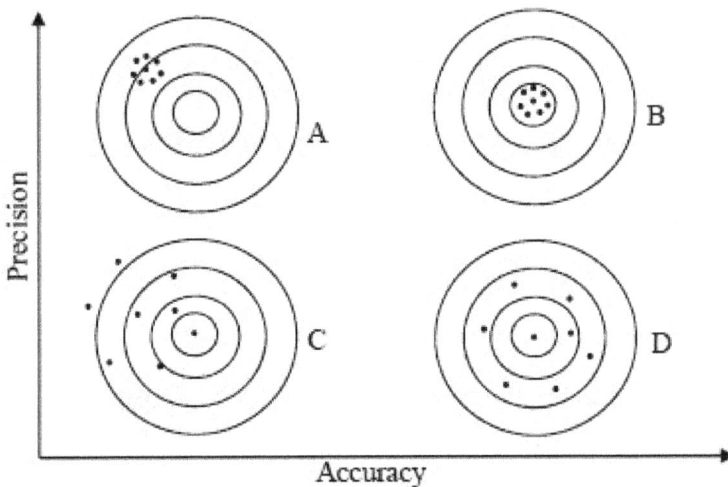

Measurement of rainfall by four rain gauges. Gauge A is precise, inaccurate; gauge B is precise, accurate; gauge C is imprecise, inaccurate; gauge D is imprecise, accurate. The innermost circle indicates the true value.

Confidence interval: The interval which includes the true value with a prescribed probability and is estimated by using the sample data.

Confidence level: The probability that the confidence interval includes the true value.

Error: The difference between the result of a measurement and the true value of the quantity measured. This term is also used for the difference between the result of a measurement and the best approximation of the true value, rather than the true value itself. The best approximation may be a mean of many measurements.

Expected value: The best approximation of the true value, which may be a mean of many measurements.

Measurement: An action intended to assign a number as the value of a physical quantity in stated units. The result of a measurement is complete if it includes an estimate of the probable magnitude of the uncertainty.

Precision: The closeness of agreement between independent measurements of a single quantity obtained by applying a stated measurement procedure several times under prescribed conditions.

Accuracy has to do with closeness to the truth, precision has to do only with closeness together.

Precision of observation or of reading is the smallest unit of division on a scale of measurement to which a reading is possible either directly or by estimation.

Random error: That part of the error that varies in an unpredictable manner, in magnitude and in sign, when measurements of the same variable are made under the same conditions.

Range: The interval between the minimum and maximum values of the quantity to be measured, for which the instrument has been constructed, adjusted or set. It can be expressed as a ratio of maximum and minimum measurable values.

References

- Davidson, E. J. (1999). "Joint application design (JAD) in practice". Journal of Systems and Software. 45 (3): 215–223. doi:10.1016/S0164-1212(98)10080-8

- Sokolowski, John A.; Banks, Catherine M., eds. (2010). Modeling and Simulation Fundamentals: Theoretical Underpinnings and Practical Domains. Hoboken, NJ: John Wiley & Sons. doi:10.1002/9780470590621. ISBN 9780470486740. OCLC 436945978

- Gemino, A.; Wand, Y. (2004). "A framework for empirical evaluation of conceptual modeling techniques". Requirements Engineering. 9 (4): 248–260. doi:10.1007/s00766-004-0204-6

- DI Spivak, RE Kent. "Ologs: a category-theoretic approach to knowledge representation" (2011). PLoS ONE (in press): e24274. doi:10.1371/journal.pone.0024274

- Slater, Matthew H.; Yudell, Zanja, eds. (2017). Metaphysics and the Philosophy of Science: New Essays. Oxford; New York: Oxford University Press. p. 127. ISBN 9780199363209. OCLC 956947667

- Gemino, A.; Wand, Y. (2003). "Evaluating modeling techniques based on models of learning". Communications of the ACM. 46 (10): 79–84. doi:10.1145/944217.944243

- Pyke, G. H. (1984). "Optimal Foraging Theory: A Critical Review". Annual Review of Ecology and Systematics. 15: 523–575

Hydrograph: Analysis, Methods and Models

A continuous graph of rate of flow versus time is known as hydrograph. Analyzing hydrography is essential for flood forecasting, flood damage reduction, and creating design flows for structures to channel floodwater. Influence of hydrography's shape and volume are affected by factors such as storm event's duration, soil types and distribution, rainfall intensity and pattern, etc. The diverse applications of hydrograph in the current scenario have been thoroughly discussed in this chapter.

Hydrograph Analysis

A hydrograph is a continuous plot of instantaneous discharge v/s time. It results from a combination of physiographic and meteorological conditions in a watershed and represents the integrated effects of climate, hydrologic losses, surface runoff, interflow, and ground water flow. A detailed analysis of hydrographs is usually important in flood damage mitigation, flood forecasting, or establishing design flows for structures that convey floodwaters.

Factors that influence the hydrograph shape and volume

    a. Meteorological factors

    b. Physiographic or watershed factors and

    c. Human factors

a. Meteorological factors include

- Rainfall intensity and pattern

- Areal distribution or rainfall over the basin and

- Size and duration of the storm event

b. Physiographic or watershed factors include

- Size and shape of the drainage area

- Slope of the land surface and main channel

- Channel morphology and drainage type

- Soil types and distribution

- Storage detention in the watershed

c. Human factors include the effects of land use and land cover

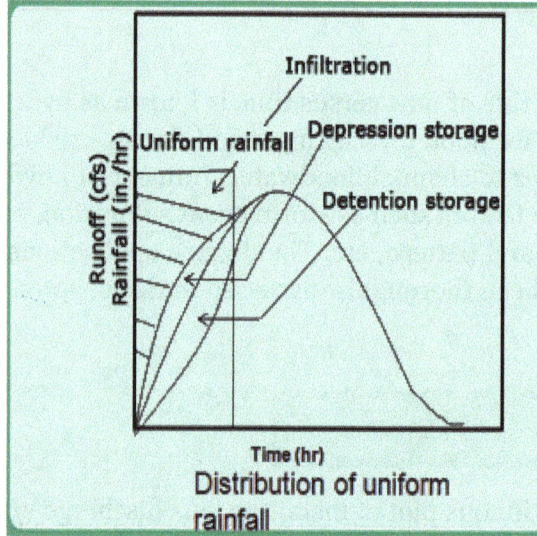

Distribution of uniform rainfall

During the rainfall, hydrologic losses such as infiltration, depression storage and detention storage must be satisfied prior to the onset of surface runoff. As the depth of surface detention increases, overland flow may occur in portion if a basin. Water eventually moves into small rivulets, small channels and finally the main stream of a watershed. Some of the water that infiltrates the soil may move laterally through upper soil zones (subsurface stromflow) until it enters a stream channel.

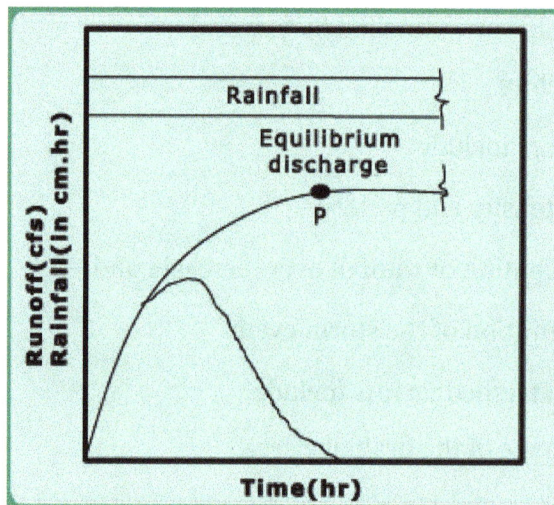

If the rainfall continues at a constant intensity for a very long period, storage is filled at some point and then an equilibrium discharge can be reached. In equilibrium discharge the inflow and outflow are equal. The point P indicates the time at which the entire discharge area contributes to the flow. The condition of equilibrium discharge is seldom observed in nature, except for very small basins, because of natural variations in rainfall intensity and duration.

Hydrograph Relations

The typical hydrograph is characterized by a

- Rising limb
- Crest
- Recession curve

The inflation point on the falling limb is often assumed to be the point where direct runoff ends.

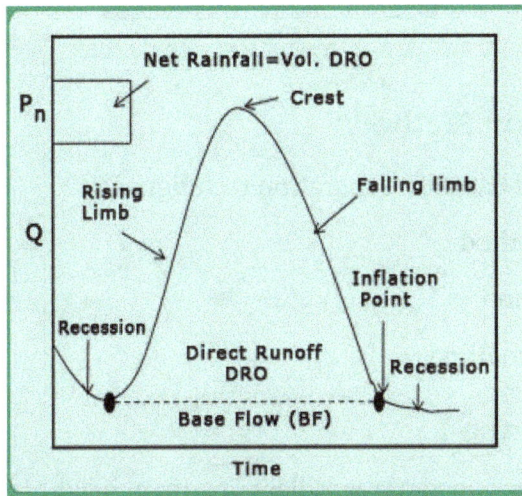

Recession and Base Flow Separation

In this the hydrograph is divided into two parts:

1. Direct runoff (DRO) and
2. Base flow (BF)

DRO includes some interflow whereas BF is considered to be mostly from contributing ground water.

Recession curve method is used to separate DRO from BF and can be expressed by an exponential depletion equation

$$q_t = q_o \cdot e^{-kt}$$

where,

q_t = discharge at a later time t

q_o = specified initial discharge

k = recession constant

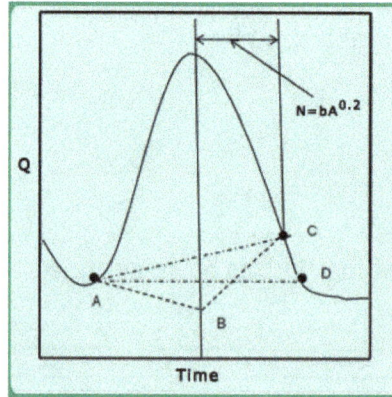

Baseflow Separation Methods

There are three types of baseflow separation techniques

1. Straight line method

2. Fixed base method

3. Constant slope method

1. Straight Line Method

- Assume baseflow constant regardless of stream height (discharge).

- Draw a horizontal line segment (A-D) from beginning of runoff to intersection with recession curve.

2. Constant Slope Method

- Connect inflection point on receding limb of storm hydrograph to beginning of storm hydrograph.

- Assumes flow from aquifers began prior to start of current storm, arbitrarily sets it to inflection point.

- Draw a line connecting the point (A-C) connecting a point N time periods after the peak.

3. Fixed Base Method

- Assume baseflow decreases while stream flow increases (i.e. to peak of storm hydrograph).

- Draw line segment (A –B) from baseflow recession to a point directly below the hydrograph peak.

- Draw line segment (B-C) connecting a point N time periods after the peak.

  where,

  N = time in days where DRO is terminated, A= Discharge area in km²,

  b= coefficient, taken as 0.827.

Rainfall Excess

The distribution of gross rainfall can be given by the continuity equation as

  Gross rainfall = depression storage+ evaporation+ infiltration + surface runoff.

In case, where depression storage is small and evaporation can be neglected, we can compute rainfall excess which equals to direct runoff, DRO, by

  Rainfall excess (Pn) = DRO = gross rainfall – (infiltration + depression storage)

The simpler method to determine rainfall excess include

  1. Horton infiltration method

  2. $\varnothing$ index method

Note:- In this, the initial loss is included for depression storage.

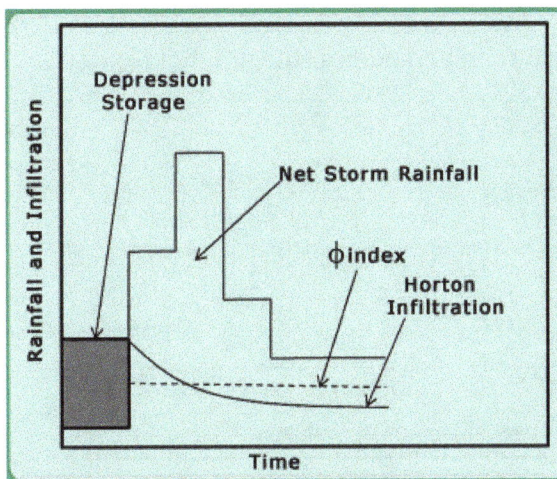

Horton Infiltration Method

Horton method estimates infiltration with an exponential-type equation that slowly declines in time as rainfall continues and is given by

$$f = f_c + (f_o - f_c)e^{-kt} \text{ (when rainfall intensity } i > f)$$

where

f = infiltration capacity (in./hr)

f_o = initial infiltration capacity (in./hr)

f_c = final infiltration capacity (in./hr)

k = empirical constant (hr^{-1})

$\emptyset$ Index Method

It is the simplest method and is calculated by finding the loss difference between gross precipitation and observed surface runoff measured as a hydrograph.

Example

Rainfall of magnitude 3.8 cm and 2.8 cm occurring on two consecutive 4-h durations on a catchment area 27 km^2 produced the following hydrograph of flow at the outlet of the catchment. Estimate the rainfall excess and ϕ-index.

Time from start of rainfall (h)	-6	0	6	12	18	24	30	36	42	48	54	60	66
Observed flow (m³/s)	6	5	13	26	21	16	12	9	7	5	5	4.5	4.5

Baseflow Separation: Using Simple straight line method,

$$N = 0.83 \, A^{0.2} = 0.83 \, (27)^{0.2}$$

$$= 1.6 \text{ days} = 38.5 \text{ h}$$

So the baseflow starts at 0th h and ends at the point (12+38.5) h

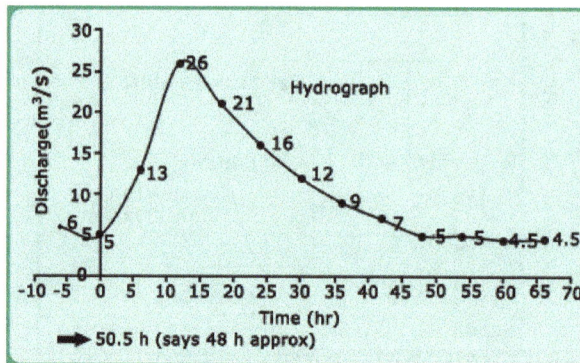

Constant baseflow of 5m³/s

Time (h)	FH Ordinates (m³/s)	DRH Ordinates (m³/s)
-6	6	1
0	5	0
6	13	8
12	26	21
18	21	16
24	16	11
30	12	7
36	9	4
42	7	2
48	5	0
54	5	0
60	4.5	0
66	4.5	0

DRH ordinates are obtained from subtracting the corresponding FH with the base flow i.e. 5 m³/s

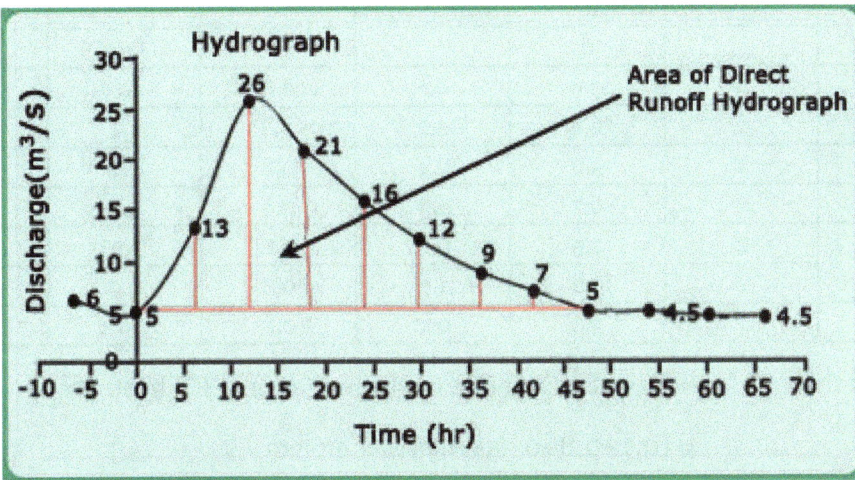

Area of DRH $= (6*60*60)[1/2 \,(8)+1/2\,(8+21) +1/2\,(21+16)+ 1/2\,(16+11)+ 1/2\,(11+7)$
$+ 1/2\,(7+4)+ 1/2\,(4+2)+ 1/2\,(2)]$

$= 1.4904 * 10^6 m^3$ (total direct runoff due to storm)

Run-off depth = Runoff volume/catchment area

$= 1.4904 * 10^6/27* 10^6$

$= 0.0552m = 5.52$ cm = rainfall excess

Total rainfall $= 3.8 +2.8 = 6.6cm$

Duration = 8h

$$\phi - \text{index} = (P - R)/t = (6.6 - 5.52)/8 = 0.135 \text{cm}/\text{h}$$

Example

A storm over a catchment of area 5.0 km² had a duration of 14 hours. The mass curve of rainfall of the storm is as follows:

Time from start of storm (h)	0	2	4	6	8	10	12	14
Observed flow (m³/s)	0	0.6	2.8	5.2	6.6	7.5	9.2	9.6

If the ϕ - index of the catchment is 0.4 cm/h, determine the effective rainfall hyetograph and the volume of direct runoff from the catchment due to the storm.

A storm over a catchment of area 5.0 km² had a duration of 14 hours. The mass curve of rainfall of the storm is as follows:

Time from start of storm (h)	Time interval Δt	Accumulated rainfall in Δt (cm)	Depth of rainfall in Δt (cm)	$\phi \Delta t$ (cm)	ER (cm)	Intensity of ER (cm/h)
0	-	0	-	-	-	-
2	2	0.6	0.6	0.8	0	0
4	2	2.8	2.2	0.8	1.4	0.7
6	2	5.2	2.4	0.8	1.6	0.8
8	2	6.7	1.5	0.8	0.7	0.35
10	2	7.5	0.8	0.8	0	0
12	2	9.2	1.7	0.8	0.9	0.45
14	2	9.6	0.4	0.8	0	0

Total effective rainfall = Direct runoff due to storm = area of ER hyetograph

$$= (0.7+0.8+0.35+0.45)^{*2} = 4.6 \text{ cm}$$

Volume of direct runoff = $(4.6/100) * 5.0*(1000)^2$

$$= 230000 \text{m}^3$$

Run-off Measurement

Time- Area Method

This method assumes that the outflow hydrograph results from pure translation of direct runoff to the outlet, at an uniform velocity, ignoring any storage effect in the watershed. The relation ship is defined by dividing a watershed into subareas with distinct runoff translation times to the outlet. The subareas are delineated with isochrones of equal translation time numbered upstream from the outlet. In a uniform rainfall intensity distribution over the watershed, water first flows from areas immediately adjacent to the outlet, and the percentage of total area contributing increases progressively in time. The surface runoff from area A_1 reaches the outlet first followed by contributions from A_2, A_3 and A_4.

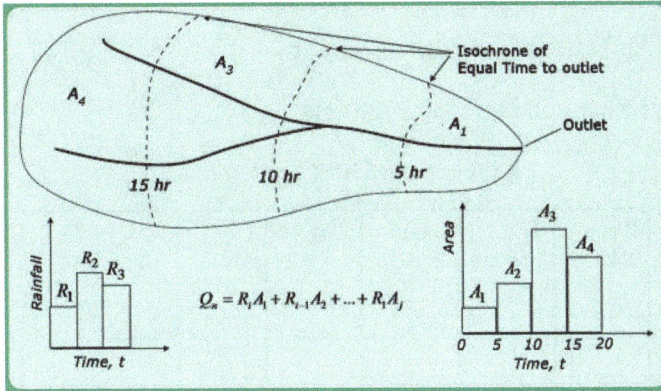

where

Q_n = hydrograph ordinate at time n (cfs)

R_i = excess rainfall ordinate at time i (cfs)

A_j = time –area histogram ordinate at time j (ft^2)

Limitation of Time Area Method

This method is limited because of the difficulty of constructing isochronal lines and the hydrograph must be further adjusted to represent storage effects in the watershed.

Example

Find the storm hydrograph for the following data using time area method. Given rainfall excess ordinate at time is 0.5 in./hr

	A	B	C	D
Area (ac)	100	200	300	100
Time to gage G (hr)	1	2	3	4

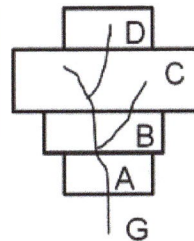

Time area histogram method uses

$$Q_n = R_i A_1 + R_{i-2} A_2 + \ldots\ldots + R_i A_j$$

For n = 5, i = 5, and j = 5

$$Q_5 = R_5 A_1 + R_4 A_2 + R_3 A_3 + R_2 A_4$$

(0.5 in./ hr) (100 ac) + (0.5 in./hr) (200 ac) + (0.5 in./hr)

(300ac) + (0.5 in./hr) (100) Q_5 = 350 ac-in./hr

Note that 1 ac-in./hr ≈ 1 cfs, hence Q_5=350 cfs

Excel spreadsheet calculation

Time (hr)	Hydro-graph ordinate (R1: Rn)	Basin No.	Time to gage	Basin Area A1 : An (ac)	R1 : An	R2 : An	R2 : An	R2 : An	R2 : An	Storm Hydro-graph
0										0
1	0.5	A	1	100	*50					50
2	0.5	B	2	200	100	50				+150
3	0.5	C	3	300	150	100	50			300
4	0.5	D	4	400	50	150	100	50		350
5						50	150	100	50	350
6							50	150	100	300
7								50	150	200
8									50	50
9										0

* =(R1*A1) = (0.5*100) and + = (adding the columns from 6 to 10)

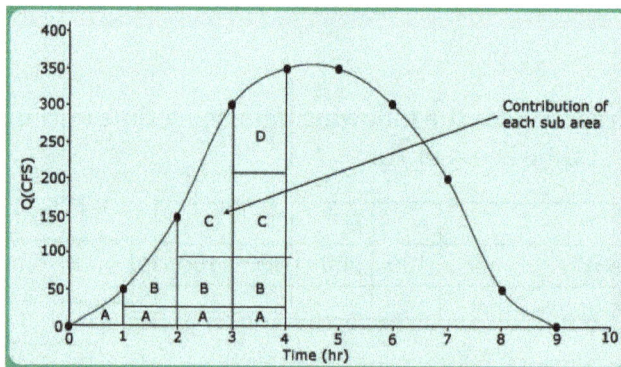

Unit Hydrograph

The unit hydrograph is the unit pulse response function of a linear hydrologic system. First proposed by Sherman (1932), the unit hydrograph (originally named unitgraph) of a watershed is defined as a direct runoff hydrograph (DRH) resulting from

1 in (usually taken as 1 cm in SI units) of excess rainfall generated uniformly over the drainage area at a constant rate for an effective duration. Sherman originally used the word "unit" to denote a unit of time. But since that time it has often been interpreted as a unit depth of excess rainfall. Sherman classified runoff into surface runoff and groundwater runoff and defined the unit hydrograph for use only with surface runoff.

The unit hydrograph is a simple linear model that can be used to derive the hydrograph resulting from any amount of excess rainfall. The following basic assumptions are inherent in this model:

- Rainfall excess of equal duration are assumed to produce hydrographs with equivalent time bases regardless of the intensity of the rain.

- Direct runoff ordinates for a storm of given duration are assumed directly proportional to rainfall excess volumes.

- The time distribution of direct runoff is assumed independent of antecedent precipitation.

- Rainfall distribution is assumed to be the same for all storms of equal duration, both spatially and temporally.

Derivation of UH : Gauged Watershed

Terminologies

1. Duration of effective rainfall : the time from start to finish of effective rainfall.

2. Lag time (L or t_p): the time from the center of mass of rainfall excess to the peak of the hydrograph.

3. Time of rise (T_R): the time from the start of rainfall excess to the peak of the hydrograph.

4. Time base (T_b): the total duration of the DRO hydrograph.

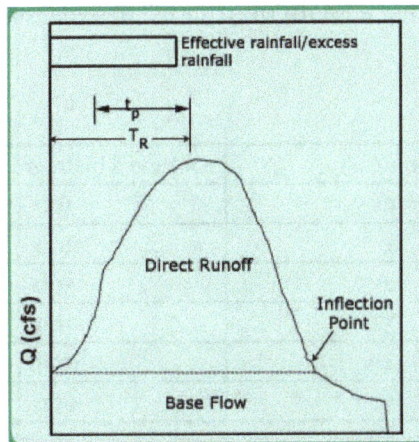

Rules to be Observed in Developing UH from Gaged Watersheds

1. Storms should be selected with a simple structure with relatively uniform spatial and temporal distributions.

2. Watershed sizes should generally fall between 1.0 and 100 mi² in modern watershed analysis.

3. Direct runoff should range 0.5 to 2 in.

4. Duration of rainfall excess D should be approximately 25% to 30% of lag time t_p.

5. A number of storms of similar duration should be analyzed to obtain an average UH for that duration.

6. Step 5 should be repeated for several rainfall of different durations.

Essential Steps for Developing UH from Single Storm Hydrograph

1. Analyze the hydrograph and separate base flow.

2. Measure the total volume of DRO under the hydrograph and convert time to inches (mm) over the watershed.

3. Convert total rainfall to rainfall excess through infiltration methods, such that rainfall excess = DRO, and evaluate duration D of the rainfall excess that produced the DRO hydrograph.

4. Divide the ordinates of the DRO hydrograph by the volume in inches (mm) and plot these results as the UH for the basin. Time base Tb is assumed constant for storms of equal duration and thus it will not change.

5. Check the volume of the UH to make sure it is 1.0 in.(1.0mm), and graphically adjust ordinates as required.

Example

Obtain a Unit Hydrograph for a basin of 315 km² of area using the rainfall and stream flow data tabulated below.

Stream Flow Data

Time (hr)	Observed hydrography (m³/s)
0	100
1	100
2	300
3	700
4	1000
5	800
6	600

7	400
8	300
9	200
10	100
11	100

Rainfall Data

Time (hr)	Gross PPT (GRH) (cm/h)
0-1	0.5
1-2	2.5
2-3	2.5
3-4	0.5

Empirical unit hydrograph derivation separates the base flow from the observed stream flow hydrograph in order to obtain the direct runoff hydrograph (DRH). For this example, use the horizontal line method to separate the base flow. From observation of the hydrograph data, the stream flow at the start of the rising limb of the hydrograph is 100 m³/s. Compute the volume of direct runoff. This volume must be equal to the volume of the effective rainfall hyetograph (ERH).

VDRH = (200+600+900+700+500+300+200+100) m³/s (3600) s = 12,600,000 m³

Express VDRH in equivalent units of depth:

V_{DRH} in equivalent units of depth = V_{DRH}/A_{basin} = 12,600,000 m³/(315000000 m²) = 0.04 m = 4 cm

Obtain a Unit Hydrograph by normalizing the DRH. Normalizing implies dividing the ordinates of the DRH by the V_{DRH} in equivalent units of depth.

Time (hr)	Observed hydrography (m³/s)	Direct Runoff Hydro-graph (DRH) (m³/s)	Unit Hydrograph (DRH) (m³/s)
0	100	0	0
1	100	0	0
2	300	200	50
3	700	600	150
4	1000	900	225
5	800	700	175
6	600	500	125
7	400	300	75
8	300	200	50
9	200	100	25
10	100	0	0
11	100	0	0

Example

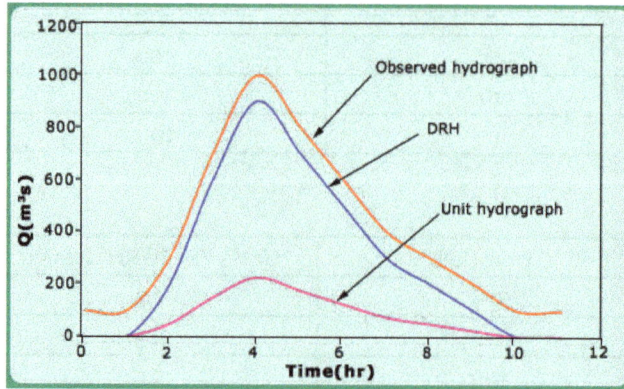

Determine the duration D of the ERH associated with the UH obtained in 4. In order to do this:

1. Determine the volume of losses, VLosses which is equal to the difference between the volume of gross rainfall, VGRH, and the volume of the direct runoff hydrograph, VDRH.

$$V_{Losses} = V_{GRH} - V_{DRH} = (0.5 + 2.5 + 2.5 + 0.5) \text{ cm/h 1 h - 4 cm = 2 cm}$$

2. Compute the f-index equal to the ratio of the volume of losses to the rainfall duration, t_r. Thus,

$$\text{ø-index} = \text{VLosses/tr} = 2 \text{ cm} / 4 \text{ h} = 0.5 \text{ cm/h}$$

3. Determine the ERH by subtracting the infiltration (e.g., ø-index) from the GRH:

Time (hr)	Effective precipitation (ERH) (cm/h)
0-1	0
1-2	2
2-3	2
3-4	0

As observed in the table, the duration of the effective rainfall hyetograph is 2 hours. Thus, D = 2 hours, and the Unit Hydrograph obtained above is a 2-hour Unit Hydrograph.

S – Curve Method

It is the hydrograph of direct surface discharge that would result from a continuous succession of unit storms producing 1cm (in.) in t_r –hr. If the time base of the unit hydrograph is T_b hr, it reaches constant outflow (Q_e) at T hr, since 1 cm of net rain on the catchment is being supplied and removed every tr hour and only T/t_r unit graphs are necessary to produce an S-curve and develop constant outflow given by,

$$Q_e = (2.78 \cdot A) / t_r$$

where

Q_e = constant outflow (cumec)

t_r = duration of the unit graph (hr)

A = area of the basin (km² or acres)

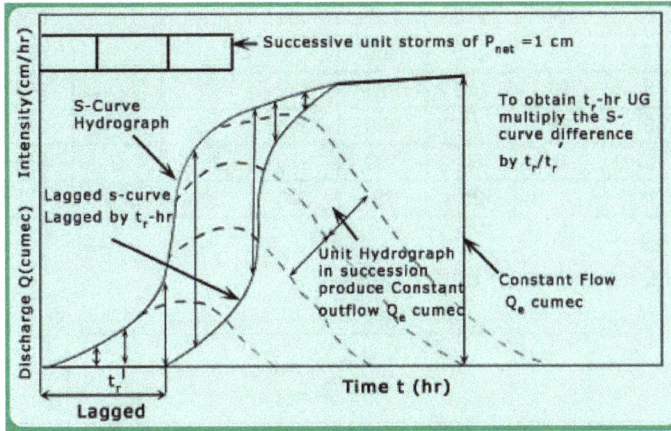

Changing the duration of UG by S-curve technique

Example

Convert the following 2-hr UH to a 3-hr UH using the S-curve method.

Time (hr)	2- hr UH ordinate (cfs)
0	0
1	75
2	250
3	300
4	275
5	200
6	100
7	75
8	50
9	25
10	0

Solution

Make a spreadsheet with the 2-hr UH ordinates, then copy them in the next column lagged by D=2 hours. Keep adding columns until the row sums are fairly constant. The sums are the ordinates of your S-curve.

Time (hr)	2-hr UH	2- hr lagged UH's					Sum
0	0						0
1	75						75
2	250	0					250
3	300	75					375
4	275	250	0				525
5	200	300	75				575
6	100	275	250	0			625
7	75	200	300	75			650
8	50	100	275	250	0		675
9	25	75	200	300	75		675
10	0	50	100	275	250	0	675
11		25	75	200	300	75	675

Draw your S-curve, as shown in figure below

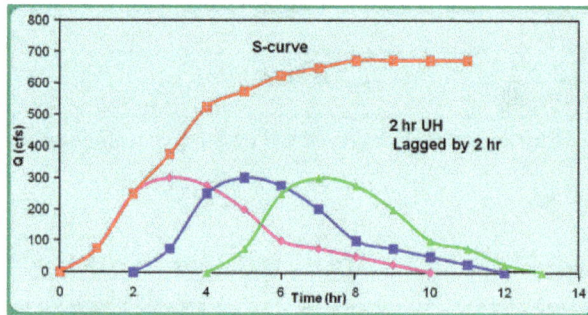

Make a spreadsheet with the 2-hr UH ordinates, then copy them in the next column lagged by D=2 hours. Keep adding columns until the row sums are fairly constant. The sums are the ordinates of your S-curve.

Time (hr)	S-curve ordinate	S-curve lagged 3hr	Difference	3- hr UH ordinate
0	0		0	0
1	75		75	50
2	250		250	166.7
3	375	0	375	250
4	525	75	450	300
5	575	250	352	216
6	625	375	250	166.7
7	650	525	125	83.3
8	675	575	100	66.7
9	675	625	50	33.3
10	675	650	25	16.7
11	675	675	0	0

Discrete Convolution Equation

Suppose that there are M pulses of excess rainfall. If N pulses of direct runoff are considered, then N equations can be written Q_n in terms of $N - M + 1$ unknown values of unit hydrograph ordinates, where $n = 1, 2, ..., N$.

$$Q_n = \sum_{m=1}^{m^*} P_m U_{n-m+1} \qquad\qquad m^* = \min(n, M)$$

where,

Q_n = Direct runoff

P_m = Excess rainfall

U_{n-m+1} = Unit hydrograph

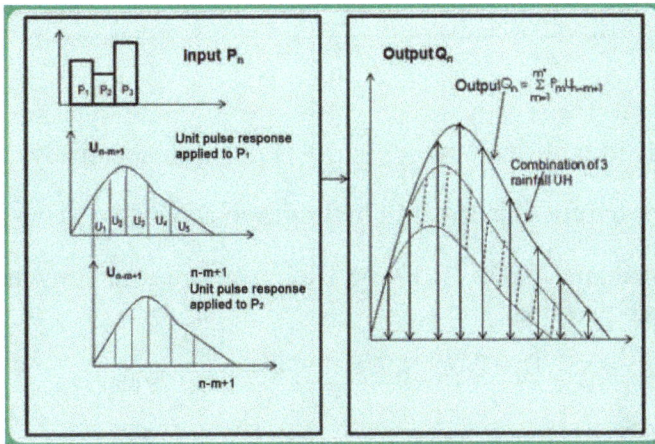

The Set of Equations for Discrete Time Convolution

$$Q_1 = P_1 U_1$$

$$Q_2 = P_2 U_1 + P_1 U_2$$

$$Q_3 = P_3 U_1 + P_2 U_2 + P_1 U_3$$

$$\vdots$$

$$Q_M = P_M U_1 + P_{M-1} U_2 + ... P_1 U_M$$

$$Q_{M+1} = 0 + P_M U_2 + P_2 U_M + ... P_1 U_{M+1}$$

$$\vdots$$

$$Q_{N-1} = 0 + 0 + ... + 0 + 0 + ... + P_M U_{N-M} + P_{M-1} U_{N-M+1}$$

$$Q_N = 0 + 0 + ... + 0 + 0 + ... + 0 + P_{M-1} U_{N-M+1}$$

$$Q_n = \sum_{m=1}^{m^*} P_m U_{n-m+1}$$

$$n = 1, 2, ..., 3$$

Example

Find the one hour unit hydrograph using the excess rainfall hyetograph and direct run-off hydrograph given in the table.

Time (1hr)	Excess Rainfall (in)	Direct Runoff (cfs)
1	1.06	428
2	1.93	1923
3	1.81	5297
4		9131
5		10625
6		7834
7		3921
8		1846
9		1402
10		830
11		313

Solution

The ERH and DRH in table have M=3 and N=11 pulses respectively.

Hence, the number of pulses in the unit hydrograph is N-M+1=11-3+1=9.

Substituting the ordinates of the ERH and DRH into the equations in table yields a set of 11 simultaneous equations.

$$U_1 = \frac{Q_2 - P_2 U_1}{P_1} = \frac{1.928 - 1.93 \times 404}{1.06} = 1,079 \text{ cfs / in}$$

Similarly calculate for remaining ordinates and the final UH is tabulated below

n	1	2	3	4	5	6	7	8	9
U_n(cfs/in)	404	1079	2343	2506	1460	453	381	274	173

Snyder's Method

Snyder (1938) was the to develop a synthetic UH based on a study of watersheds in the Appalachian Highlands. In basins ranging from 10 – 10,000 mi².

Snyder relations are

$$t_p = C_t (LL_C)^{0.3}$$

where,

t_p = basin lag (hr)

L = length of the main stream from the outlet to the divide (mi)

L_c = length along the main stream to a point nearest the watershed centroid (mi)

C_t = Coefficient usually ranging from 1.8 to 2.2

$$Q_p = 640 \, C_p A / t_p$$

where,

Q_p = peak discharge of the UH (cfs)

A = Drainage area (mi²)

C_p = storage coefficient ranging from 0.4 to 0.8, where larger values of cp are associated with smaller values of Ct

$$T_b = 3 + t_p / 8$$

where,

T_b is the time base of hydrograph

Note: For small watershed the above eq. should be replaced by multiplying t_p by the value varies from 3-5.

The above 3 equations define points for a UH produced by an excess rainfall of duration

$$D = t_p / 5.5$$

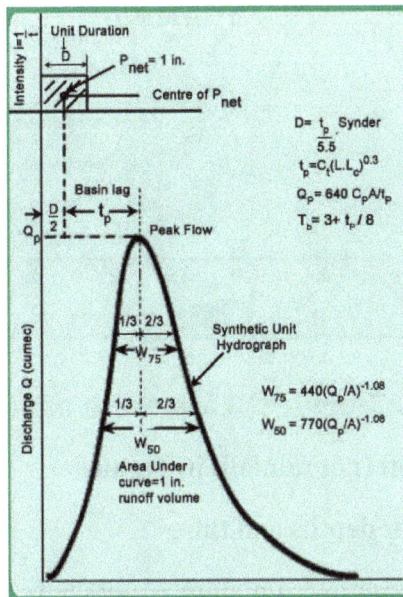

Snyder's hydrograph parameter

Use Snyder's method to develop a UH for the area of 100mi² described below. Sketch the appropriate shape. What duration rainfall does this correspond to?

$$C_t = 1.8, \quad L = 18\text{mi},$$

$$C_p = 0.6, \quad L_C = 10\text{m}$$

Calculate t_p

$$t_p = C_t(LL_c)0.3$$

$$= 1.8(18 \cdot 10)0.3\text{hr},$$

$$= 8.6\text{hr}$$

Since this is a small watershed,

$$T_b \approx 4t_p = 4(8.6)$$

$$= 34.4\text{hr}$$

Calculate Q_p

$$Q_p = 640(c_p)(A)/t_p$$

$$= 640(0.6)(100)/8.6$$

$$= 4465 \text{ cfs}$$

Duration of rainfall

$$D = t_p/5.5\text{hr}$$

$$= 8.6/5.5\text{hr}$$

$$= 1.6\text{hr}$$

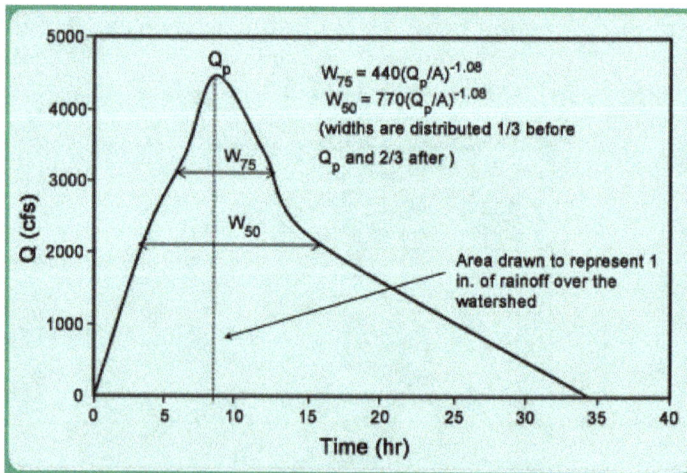

SCS (Soil Conservation Service) Unit Hydrograph

- Unit = 1 inch of runoff (not rainfall) in 1 hour.
- Can be scaled to other depths and times.
- Based on unit hydrographs from many watersheds.

The earliest method assumed a hydrograph as a simple triangle, with rainfall duration D, time of rise T_R (hr), time of fall B and peak flow Q_p (cfs).

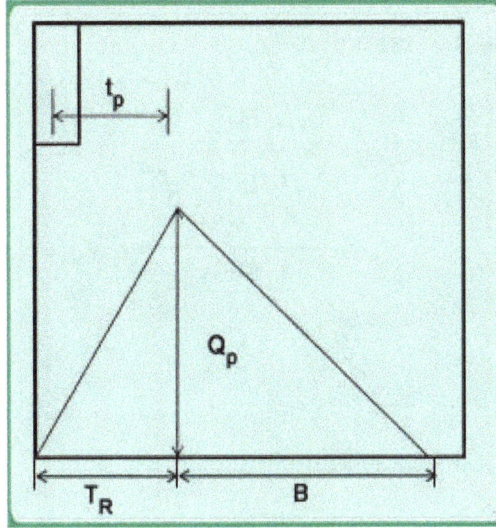

SCS triangular UH

The volume of direct runoff is

$$Vol = \frac{Q_p T_R}{2} + \frac{Q_p B}{2} \quad \text{or} \quad Q_p = \frac{2vol}{T_R + B}$$

where B is given by

$$B = 1.67 T_R$$

Therefore runoff eq. becomes, for 1 in. of rainfall excess,

$$Q_p = \frac{0.75vol}{T_R} = Q_p = \frac{0.75(640)A(1.008)}{T_R}$$

$$Q_p = \frac{484A}{T_R}$$

where,

A = area of basin (sq mi)

T_R = time of rise (hr)

Time of rise T_R is given by

$$T_R = \frac{D}{2} + t_p$$

where,

D = rainfall duration (hr)

t_p = lag time from centroid of rainfall to Q_p

Lag time is given by

$$t_p = \frac{L^{0.8}\left(\dfrac{1000}{CN} - 9\right)^{0.7}}{19000\,y^{0.5}}$$

where,

L = length to divide (ft)

Y = average watershed slope (in present)

CN = curve number for various soil/land use

Runoff curve number for different land use

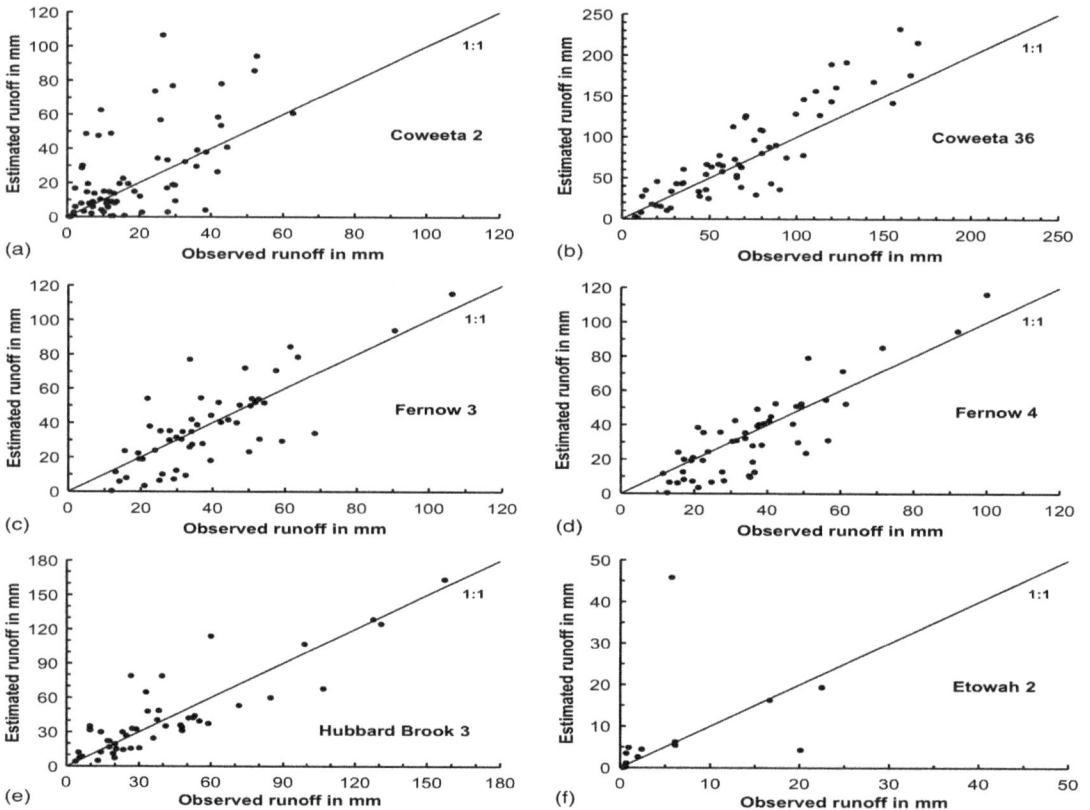

Example

Use the SCS method to develop a UH for the area of 10 mi² described below. Use rainfall duration of D = 2 hr.

$$C_p = 1.8, \quad L = 5\text{mi},$$

$$C_p = 0.6, \quad L_c = 2\text{mi}$$

The watershed consist CN = 78 and the average slope in the watershed is 100 ft/mi. Sketch the resulting SCS triangular hydrograph.

Solution

Find t_p by the eq.

$$t_p = \frac{L^{0.8}\left(\dfrac{1000}{CN} - 9\right)^{0.7}}{19000 y^{0.5}}$$

Convert L= 5 mi, or (5*5280 ft/mi) = 26400 ft.

Slope is 100 ft/mi, so y = (100 ft/mi) (1 mi/5280 ft)(100%) = 1.9%

Substituting these values in eq. of t_p, we get $t_p = 3.36$ hr

Find T_R using eq.

$$T_R = \frac{D}{2} + t_p$$

Given Rainfall duration is 2 hr, $T_R = 4.36$ hr, the rise of the Hydrograph. Then find Q_p using the eq., given $A = 10 \text{ mi}^2$

$$Q_p = \frac{484\,A}{T_R}. \text{ Hence } Q_p = 1.110 \text{ cfs}$$

To complete the graph, it is also necessary to know the time of fall B. The volume is known to be 1 in. of direct runoff over the watershed.

So, Vol=(10 mi²) (5280 ft/mi)² (ac/43560 ft²) (1 in.) = 6400 ac-in

Hence from eq.

$$Vol = \frac{Q_p T_R}{2} + \frac{Q_p B}{2}$$

$$B = 7.17\text{hr}$$

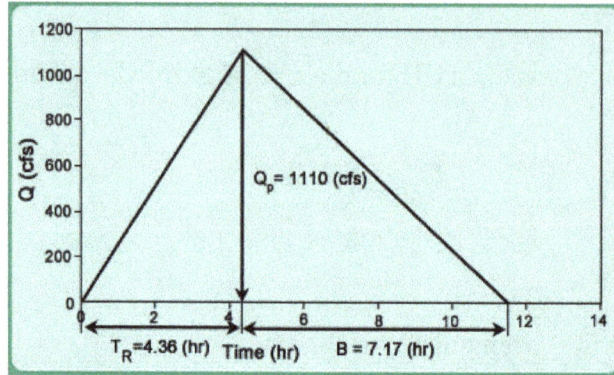

Example

1. The stream flows due to three successive storms of 2.9, 4.9 and 3.9 cm of 6 hours duration each on a basin are given below. The area of the basin is 118.8 km2 . Assuming a constant base flow of 20 cumec, derive a 6-hour unit hydrograph for the basin. An average storm loss of 0.15 cm/hr can be assumed. (Hint:- Use UH convolution method).

Time (hr)	0	3	6	9	12	15	18	21	24	27	30	33
Flow (cumec)	20	50	92	140	199	202	204	144	84	45	29	20

2. The ordinates of a 4-hour unit hydrograph for a particular basin are given below. Derive the ordinates of (i) the S-curve hydrograph, and (ii) the 2-hour unit hydrograph, and plot them, area of the basin is 630 km².

Time (hr)	Discharge (cumec)	Time (hr)	Discharge (cumec)
0	0	14	70
2	25	16	30
4	100	18	20
6	160	20	6
8	190	22	1.5
10	170	24	0
12	110		

3. The following are the ordinates of the 9-hour unit hydrograph for the entire catchment of the river Damodar up to Tenughat dam site: and the catchment characteristics are, A = 4480 km², L = 318 km, L_{ca} = 198 km. Derive a 3-hour unit hydrograph for the catchment area of river Damodar up to the head of Tenughat reservoir, given the catchment characteristics as, A = 3780 km², L = 284 km, L_{ca} = 184km. Use Snyder's approach with necessary modifications for the shape of the hydrograph.

Time (hr)	0	9	18	27	36	45	54	63	72	81	90
Flow (cumec)	0	69	1000	210	118	74	46	26	13	4	0

Kinematic Wave

In gravity and pressure driven fluid dynamical and geophysical mass flows such as ocean waves, avalanches, debris flows, mud flows, flash floods, etc., kinematic waves are important mathematical tools to understand the basic features of the associated wave phenomena. These waves are also applied to model the motion of highway traffic flows.

In these flows, mass and momentum equations can be combined to yield a kinematic wave equation. Depending on the flow configurations, the kinematic wave can be linear or non-linear, which depends on whether the wave celerity is a constant or a variable. Kinematic wave can be described by a simple partial differential equation with a single unknown field variable (e.g., the flow or wave height, h) in terms of the two independent variables, namely the time (t) and the space (x) with some parameters (coefficients) containing information about the physics and geometry of the flow. In general, the wave can be advecting and diffusing. However, in simple situation, the kinematic wave is mainly advecting.

Kinematic Wave for Debris Flow

Non-linear kinematic wave for debris flow can be written as follows with complex non-linear coefficients:

$$\frac{\partial h}{\partial t} + C \frac{\partial h}{\partial x} = D \frac{\partial^2 h}{\partial x^2},$$

where h is the debris flow height, t is the time, x is the downstream channel position, C is the pressure gradient and the depth dependent nonlinear variable wave speed, and D is a flow height and pressure gradient dependent variable diffusion term. This equation can also be written in the conservative form:

$$\frac{\partial h}{\partial t} + \frac{\partial F}{\partial x} = 0,$$

where F is the generalized flux that depends on several physical and geometrical parameters of the flow, flow height and the hydraulic pressure gradient. For $F = h^2/2$, this equation reduces to the Burgers' equation.

Kinematic Overland Flow Routing

For the conditions of kinematic flow, and with no appreciable backwater effect, the discharge can be described as a function of area only, for all x and t;

$$Q = \alpha \cdot A^m$$

where,

Q = discharge in cfs

A = cross-sectional area

α, m = kinematic wave routing parameters

Governing Equations

Henderson (1966) normalized momentum equation in the form of

$$Q = Q_o \left(1 - \frac{1}{S_o} \left(\frac{\partial y}{\partial x} + \frac{v \partial v}{g \partial x} + \frac{1 \partial v}{g \partial t} + \frac{qv}{gy} \right) \right)^{\frac{1}{2}}$$

where Qo = flow under uniform condition

Less than one, then the equation will represent Kinematic flow

Hence, for the kinematic flow condition,

$$Q \approx Q_o$$

Woolhiser and Liggett (1967) analyzed characteristics of the rising overland flow hydrograph and found that the dynamic terms can generally be neglected if,

$$k = \frac{S_o L}{y F r^2} \geq 10 \ \ or \ k = \frac{S_o L g}{v^2} \geq 10$$

where,

L = length of the plane

F_r = Froude number

y = depth at the end of the plane

S_o = slope

k = dimensionless kinematic flow number

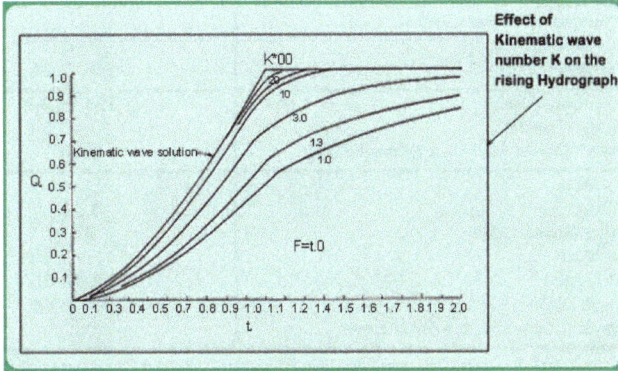

Q* is the dimensionless flow v/s t* (dimensionless time) for varies values of k in equation. It can be seen that for k≤10, large errors in calculation of Q* result by deleting dynamic terms from the momentum eq. for overland flow.

The momentum eq. for an overland flow segment on a wide plane with shallow flows can be derived from eq. and Manning's eq. for overland flow

$$q = \frac{k_m}{n}\sqrt{S_0}\, y^{5/3} \tag{a}$$

Rewriting the above equation in terms of flow per unit width for an overland flow q_0, we have

$$q_0 = \alpha_0 y_0^{m_0} \tag{b}$$

$$\alpha_0 = \frac{k_m}{n}\sqrt{S_0} = \text{conveyance factor}$$

m_o = 5/3 from manning's eq.

S_o = Average overland flow slope

y_o = mean depth of overland flow

Estimates of Manning's Roughness Coefficients for Overland Flow

Surface	N value
Asphalt/Concrete*	0.05 - 0.15
Bare Packed Soil Free of Stone	0.10
Fallow – No Residue	0.008 - 0.012
Convential Tillage – No Residue	0.06 - 0.12
Convential Tillage – With Residue	0.16 - 0.22
Chisel Plow – No Residue	0.06 - 0.12
Chisel Plow – With Residue	0.10 - 0.16
Fall Disking – With Residue	0.30 - 0.50
No Till – No Residue	0.04 - 0.10
No Till (20-40 percent residue cover)	0.07 - 0.17
No Till (60-100 percent residue cover)	0.17 - 0.47

Sparse Rangeland with Debris:	
0 Percent Cover	0.09 - 0.34
20 Percent Cover	0.05 - 0.25
Sparse Vegetation	0.053 - 0.13
Short Grass Prarie	0.10 - 0.20
Poor Grass Cover on Moderately Rough	0.30
Bare Surface	
Light Turf	0.20
Average Grass Cover	0.4
Dense Turf	0.17 - 0.80
Dense Grass	0.17 - 0.30
Bermuda Grass	0.30 - 0.48
Dense Shrubbery and Forest Litter	0.4

Kinematic Routing Methods

The continuity Eq. is

$$\frac{\partial y_o}{\partial t} + \frac{\partial q_o}{\partial x} = i - f \tag{c}$$

where,

i = rate of gross rainfall (ft/s)

f = infiltration rate

q_o = flow per unit width (cfs/ft)

y_o = mean depth of overland flow

Finally, by substituting equation c in equation a, we have

$$\frac{\partial y_0}{\partial t} + \alpha_0 m_0 y_0^{m_0 - 1} \frac{\partial y_0}{\partial x} = i - f \tag{d}$$

Equation (b) and equation (d) form the complete kinematic wave equation for overland flow.

Kinematic Channel Modeling

Representative of collectors or stream channels

1. Triangular

2. Rectangular

3. Trapezoidal

4. Circular

These are completely characterized by slope, length, cross-sectional dimensions, shape and Manning's n value.

Basic Channel Shapes and their Variations

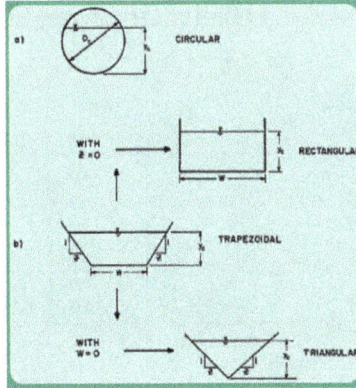

Equations of Kinematic Channel Modeling

The basic forms of the equations are similar to the overland flow Eq. (Eqs c and d). For stream channels or collectors,

$$\frac{\partial A_c}{\partial t} + \frac{\partial Q_c}{\partial x} = q_0$$

$$Q_c = \alpha_c A_c^{m_c}$$

where,

A_c = cross sectional flow area (ft^2)

Q_c = discharge

q_0 = overland inflow per unit length (cfs/ft)

α_c, m_c = kinematic wave parameter for the particular channel

Kinematic Channel Parameters

shape	α_c	m_c
Triangular	$\dfrac{0.94\sqrt{s}}{n}\left(\dfrac{z}{1+z^2}\right)^{1/3}$	4/3
Square	$\dfrac{0.72\sqrt{s}}{n}$	4/3
Rectangular	$\dfrac{1.49\sqrt{s}}{n}\left(W^{-2/3}\right)$	5/3
Trapezoidal	Variable, function of A and W	
Circular	$\dfrac{0.804\sqrt{s}}{n}\left(D_c^{1/6}\right)$	5/4

Example

Determine α_c and m_c for the case of a triangular prismatic channel

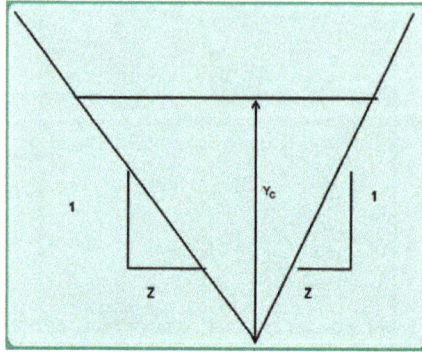

Solution

$Area = A_c = Zy_c^2$ and y_c = channel depth

Wetted perimeter = $P_c = 2y_c\sqrt{1+z^2}$

hydraulic radius = $R = \dfrac{A_c}{P_c}$

Substituting these into manning's Eq. given by

$$Q_c = \frac{1.49}{n}\sqrt{s}\,\frac{A_c^{5/3}}{P_c^{2/3}}$$

$$Q_c = \frac{1.49}{n}\sqrt{s}\,\frac{(Z^{5/3}y_c^{10/3})}{1.59y_c^{2/3}(1+Z^2)^{1/3}}$$

$$Q_c = \frac{0.94}{n}\sqrt{s}\left(\frac{Z}{1+Z^2}\right)^{1/3}(Zy_c^2)^{4/3}$$

$$Q_c = \frac{0.94}{n}\sqrt{s}\left(\frac{Z}{1+Z^2}\right)^{1/3}(A_c)^{4/3}$$

From equation, $Q_c = \alpha_c A_c^m$, Therefore,

$$\alpha_c = \frac{0.94}{n}\sqrt{s}\left(\frac{Z}{1+Z^2}\right)^{1/3} \text{ and } m_c = 4/3$$

Statistical Evaluation in Hydrology

Statistical analysis is preferred to observing hydrological variables by applying physical laws. The reason is that the data received is inadequate, plausibly incorrect, and inherits randomness. Statistics also allows for probability and adjustability to real-world situations. This chapter is an overview of the subject matter incorporating all the major aspects of hydrology.

Statistical Analysis

A hydrologic process is a phenomenon describing the occurrence and movement of water in the earth phase of the hydrologic cycle. When the phenomena are observed at intervals of time, it is termed as a discrete process and when these are observed continuously, continuous series are generated. The observed behavior of many hydrologic variables cannot be completely explained by applying the physical laws. Three reasons behind this are: a) incorrect or incomplete knowledge of the underlying processes, b) inadequate data, and c) inherent randomness of the process. In these situations we resort to statistical analysis to infer and predict the behavior of the hydrologic variables.

Data of the concerned process are collected to apply the probability theory. All possible observations of a process constitute its population. But we cannot completely collect the data of population in a limited time frame. Generally, only a finite portion of the population is observed and is called a sample. Statistical properties of the sample are determined and it is assumed that these represent the properties of the population.

Many hydrologic data (e.g., daily rainfall amount, discharge of a river) can be characterized by variable(s) that are unpredictable to some degree. Yet frequently there is a degree of consistency in the factors governing the outcome which exhibits a statistical regularity. A variable whose value at any time is not influenced by the value at earlier time(s) is known as a *random variable*. A random variable can be discrete which means that it can take on only a finite set of values, for example the number of rainy days in a year at Roorkee. It can also be continuous (a time series) and can take on any value, for example, the water level of Ganga River at Rishikesh or the magnitude of rainfall at a place.

The behavior of the hydrological processes may be deterministic or stochastic. For a deterministic process, the behavior can be completely described with certainty. The governing equation defines the process for the entire time (or space) domain. In contrast to this, a stochastic process evolves, entirely or in major part, according to a random mechanism and hence the future outcomes of the process are not exactly predictable. Such hydrologic variables whose values are governed by the laws of chance are called stochastic variables.

For many hydrological problems in these cases, sample data consist of measurements made on a single random variable and the techniques of analysis are called univariate analysis and estimation. Univariate analysis is carried out by using the measurements of the random variable (or the sample information) to identify the statistical properties of the population from which the sample measurements are likely to have come. After the underlying population has been identified, one can make probabilistic statements about the future occurrences of the random variable; this represents univariate estimation.

Before commencing the analysis, it is always useful to plot the sample data. The following are the main steps of statistical analysis of data:

i)   Compute the basic statistical indicators of the data.

ii) Select a set of probability distribution functions.

iii) Fit the selected distributions with the sample data. Common methods of parameter estimation are discussed subsequently.

iv) Select the best fit distribution using the goodness-of-fit tests.

v)   Use the best-fit probability distribution to make inferences about the likelihood of occurrence of the magnitudes of the random variable.

If all the values of a random variable and the corresponding probabilities are known or found, the relation between these values and probabilities is described by a probability distribution. Knowing this distribution, the probability of any value of the random variable can be determined.

Probability

Based on the daily experiences, most people have an intuitive appreciation of the concept of probability or chance. When a coin is thrown, there are two equally likely possible outcomes: head or tail. Let the coin be tossed n times and head occurs s times. The ratio s/n is the probability of occurrence of heads.

Sample space is a collection of all possible random events that might arise from a random experiment. Sample space S and two sets of events A and B are shown in figure.

If two events A and B do not contain any common sample points, they are said to be mutually exclusive.

If two events A and B are not mutually exclusive, the common set is called their *intersection*, denoted as $A \cap B$. The union of two events A and B gives the event which is the collection of all sample points occurring at least once in either A or B and is denoted as $A \cup B$. Figure shows this concept through the Venn diagram. The complement A^c of an event A consists of all sample points in the sample space of the experiment not included in the event A.

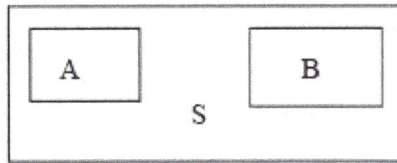

Two mutually exclusive events A and B in sample space S.

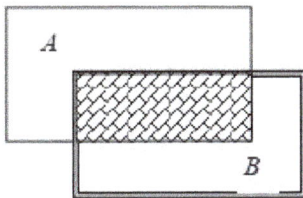
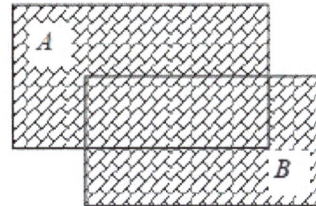

$A \cap B$ $A \cup B$

Venn diagram showing intersection and union of two events.

Axioms of Probabilities

Events are basically sets and the common set operators, including union, intersection, and complement, are applicable to them. Four frequently used rules for set operations, considering three events A, B and C are:

1. Commutative rule: $A \cup B = B \cup A$, $A \cap B = B \cap A$

2. Associative rule: $(A \cup B) \cup C = A \cup (B \cup C)$, $(A \cap B) \cap C = A \cap (B \cap C)$

3. Distributive rule: $A \cap (B \cup C) = (A \cap B) \cup (A \cap C)$, $A \cup (B \cap C) = (A \cup B) \cap (A \cup C)$

4. De Morgan's rule: $(A \cup B)^c = A^c \cap B^c$, $(A \cup B)^c = A^c \cap B^c$

The notation $P[A]$ is used to denote the probability of a random event A. Now we discuss the axioms of probability.

Axiom 1: The probability of an event A is a number greater than or equal to zero but less than or equal to unity:

$$0 \le P[A] \le 1$$

Axiom 2: The probability of an event A, whose occurrence is certain, is unity:

$$P[A] = 1$$

where A is the event associated with all sample points in the sample space.

Axiom 3: The probability of an event which is the union of two events is:

$$P[A \ or \ B] = P[A \cup B] = P[A] + P[B] - P[A \cap B]$$

where $A \cup B$ denotes the union of events A and B which means that either event A occurs or event B occurs, and $A \cap B$ denotes the intersection of event A and event B. Eq. can be extended to the union of n events. If A and B are two mutually exclusive (disjointed) events, the probability of A and B, $P[A \cap B]$, will be zero and eq. becomes:

$$P[A \ or \ B] = P[A] + P[B]$$

Axiom 4: The probability of two (statistically) independent events occurring simultaneously or in succession is the product of individual probabilities:

$$P[E_1 \cap E_2] = P[E_1] \times P[E_2]$$

Statistical independence implies that the occurrence of event E_1 has no influence on the occurrence of event E_2.

Properties of Random Variable

Let X denote a random variable and x be a possible value of X. The cumulative distribution function (CDF), $F_X(x)$ is the probability that the random variable X is less than or equal to x:

$$F_X(x) = P(X \le x)$$

The probability distribution function (PDF) describes the relative likelihood that a continuous random variable X takes on different values, and is the derivative of the CDF:

$$f_X(x) = d\{F_X(x)\} / dx$$

The PDF and CDF of a random variable are shown in figure. Note that the CDF is denoted by capital letter F and the PDF by lower case letter f.

Let us now state some important properties of $f(x)$ and $F(x)$ for continuous and discrete random variables.

1. The probability of a random variable cannot be negative

$$f(x) \geq 0, \quad -\infty < x < \infty$$

2. The sum of probabilities of all possible outcomes is equal to 1, i.e., the area under the PDF is unity.

$$\int_{-\infty}^{\infty} f(x)dx = 1$$

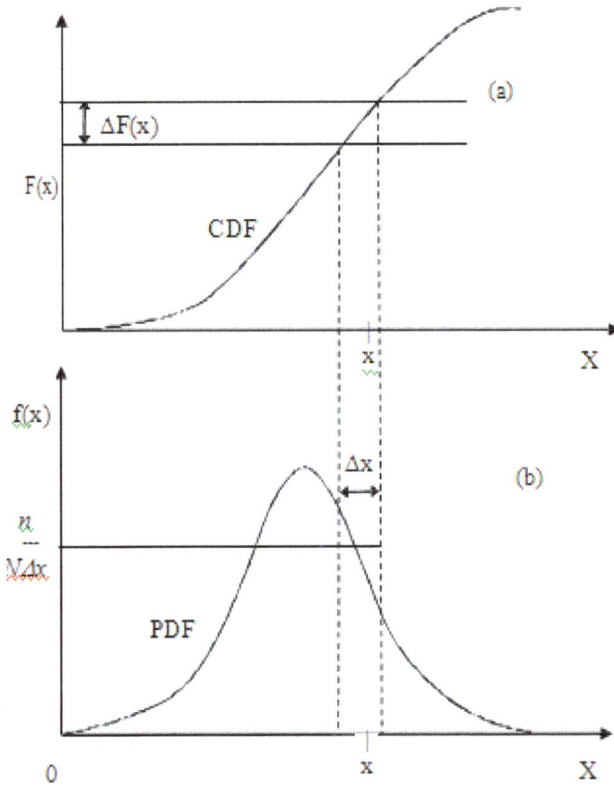

PDF and CDF of a random variable.

For discrete random variables, the following statements can be made:

$$\sum_{i} f(x) = 1$$

where $f(x_i)$ represents the probability of $X = x_i$ in the sample space if the sample contains finite and $f(x_i)$ observations can be replaced by $p(x_i)$. The probability of "X assumes a value $\leq x$" is equal to the area under PDF curve to the left of x:

3. $P(X \leq x) = F(X \leq x) = F(x) = \int_{-\infty}^{x} f(x)dx$

Moreover

$$P(a < X \le b) = P(X \le b) - P(X \le a)$$

$$= \int_{-\infty}^{b} f(x)dx - \int_{-\infty}^{a} f(x)dx$$

$$= \int_{a}^{b} f(x)dx, \quad for\ a < b$$

For discrete case

$$P(a \le x \le b) = \sum_{x_i \ge a}^{x_i \le b} p(x_i)$$

$$P(X \le x_k) = \sum_{i=1}^{k} p(x_i)$$

Probability Distribution

In probability theory and statistics, a probability distribution is a mathematical function that, stated in simple terms, can be thought of as providing the probability of occurrence of different possible outcomes in an experiment. For instance, if the random variable X is used to denote the outcome of a coin toss ('the experiment'), then the probability distribution of X would take the value 0.5 for $X = \text{heads}$, and 0.5 for $X = \text{tails}$. .

In more technical terms, the probability distribution is a description of a random phenomenon in terms of the probabilities of events. Examples of random phenomena can include the results of an experiment or survey. A probability distribution is defined in terms of an underlying sample space, which is the set of all possible outcomes of the random phenomenon being observed. The sample space may be the set of real numbers or a higher-dimensional vector space, or it may be a list of non-numerical values; for example, the sample space of a coin flip would be $\{\text{heads}, \text{tails}\}$.

Probability distributions are generally divided into two classes. A discrete probability distribution (applicable to the scenario where the set of possible outcomes is discrete, such as a coin toss or a roll of dice) can be encoded by a discrete list of the probabilities of the outcomes, known as a probability mass function. On the other hand, a continuous probability distribution (applicable to the scenarios where the set of possible outcomes can take on values in a continuous range (e.g., real numbers), such as the temperature on a given day) is typically described by probability density functions (with the probability of any individual outcome actually being 0). The normal distribution represents a commonly encountered continuous probability distribution. More complex experiments, such as those involving stochastic processes defined in continuous time, may demand the use of more general probability measures.

A probability distribution whose sample space is the set of real numbers is called univariate, while a distribution whose sample space is a vector space is called multivariate. A univariate distribution gives the probabilities of a single random variable taking on various alternative values; a multivariate distribution (a joint probability distribution) gives the probabilities of a random vector—a list of two or more random variables—taking on various combinations of values. Important and commonly encountered univariate probability distributions include the binomial distribution, the hypergeometric distribution, and the normal distribution. The multivariate normal distribution is a commonly encountered multivariate distribution.

Introduction

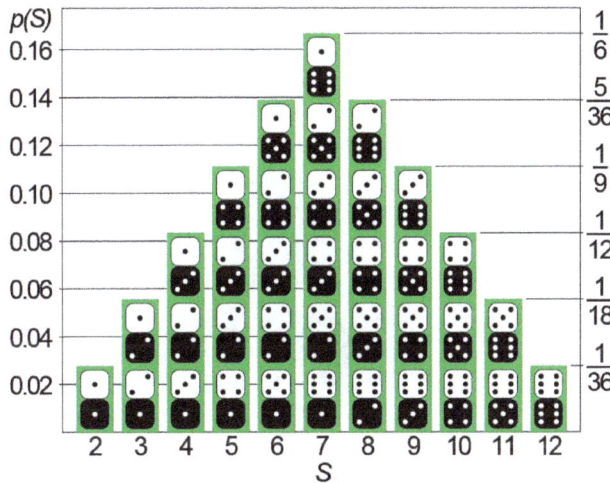

The probability mass function (pmf) $p(S)$ specifies the probability distribution for the sum S of counts from two dice. For example, the figure shows that $p(11) = 1/18$. The pmf allows the computation of probabilities of events such as $P(S > 9) = 1/12 + 1/18 + 1/36 = 1/6$, and all other probabilities in the distribution.

To define probability distributions for the simplest cases, one needs to distinguish between discrete and continuous random variables. In the discrete case, it is sufficient to specify a probability mass function assigning a probability to each possible outcome: for example, when throwing a fair dice, each of the six values 1 to 6 has the probability $1/6$. The probability of an event is then defined to be the sum of the probabilities of the outcomes that satisfy the event; for example, the probability of the event "the die rolls an even value" is

$$\text{Prob}(2) + \text{Prob}(4) + \text{Prob}(6) = 1/6 + 1/6 + 1/6 = 1/2.$$

In contrast, when a random variable takes values from a continuum then typically, any individual outcome has probability zero and only events that include infinitely many outcomes, such as intervals, can have positive probability. For example, the probability that a given object weighs *exactly* 500 g is zero, because the probability of measuring exactly 500 g tends to zero as the accuracy of our measuring instruments increases. Nevertheless, in quality control one might demand that the probability of a "500 g" package containing between 490 g and 510 g should be no less than 98%, and this demand is less sensitive to the accuracy of our instruments.

Continuous probability distributions can be described in several ways. The probability density function describes the infinitesimal probability of any given value, and the probability that the outcome lies in a given interval can be computed by integrating the probability density function over that interval. On the other hand, the cumulative distribution function describes the probability that the random variable is no larger than a given value; the probability that the outcome lies in a given interval can be computed by taking the difference between the values of the cumulative distribution function at the endpoints of the interval. The cumulative distribution function is the antiderivative of the probability density function provided that the latter function exists.

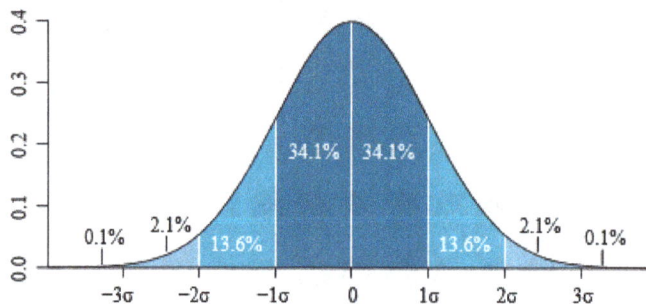

The probability density function (pdf) of the normal distribution, also called Gaussian or "bell curve", the most important continuous random distribution. As notated on the figure, the probabilities of intervals of values correspond to the area under the curve.

Terminology

As probability theory is used in quite diverse applications, terminology is not uniform and sometimes confusing. The following terms are used for non-cumulative probability distribution functions:

- Distribution, Frequency distribution: is a table that displays the frequency of various outcomes in a sample.

- Probability distribution: is a table that displays the probabilities of various outcomes in a sample. Could be called a "normalized frequency distribution table", where all occurrences of outcomes sum to 1.

- Distribution function: is a functional form of frequency distribution table.

- Probability distribution function: is a functional form of probability distribution table. Could be called a "normalized frequency distribution function", where area under the graph equals to 1.

Finally,

- Probability mass, Probability mass function, p.m.f., Discrete probability distribution function: for discrete random variables.

- Categorical distribution: for discrete random variables with a finite set of values.

- Probability density, Probability density function, p.d.f., Continuous probability distribution function: most often reserved for continuous random variables.

The following terms are somewhat ambiguous as they can refer to non-cumulative or cumulative distributions, depending on authors' preferences:

- Probability distribution function: continuous or discrete, non-cumulative or cumulative.

- Probability function: even more ambiguous, can mean any of the above or other things.

Basic Terms

- Mode: for a discrete random variable, the value with highest probability (the location at which the probability mass function has its peak); for a continuous random variable, the location at which the probability density function has its peak.

- Support: the smallest closed set whose complement has probability zero.

- Head: the range of values where the pmf or pdf is relatively high.

- Tail: the complement of the head within the support; the large set of values where the pmf or pdf is relatively low.

- Expected value or mean: the weighted average of the possible values, using their probabilities as their weights; or the continuous analog thereof.

- Median: the value such that the set of values less than the median has a probability of one-half.

- Variance: the second moment of the pmf or pdf about the mean; an important measure of the dispersion of the distribution.

- Standard deviation: the square root of the variance, and hence another measure of dispersion.

- Symmetry: a property of some distributions in which the portion of the distribution to the left of a specific value is a mirror image of the portion to its right.

- Skewness: a measure of the extent to which a pmf or pdf "leans" to one side of its mean. The third standardized moment of the distribution.

- Kurtosis: a measure of the "fatness" of the tails of a pmf or pdf. The fourth standardized moment of the distribution.

Cumulative Distribution Function

Because a probability distribution Pr on the real line is determined by the probability of a scalar random variable X being in a half-open interval $(-\infty, x]$, the probability distribution is completely characterized by its cumulative distribution function:

$$F(x) = \Pr[X \le x] \qquad \text{for all } x \in \mathbb{R}.$$

Discrete Probability Distribution

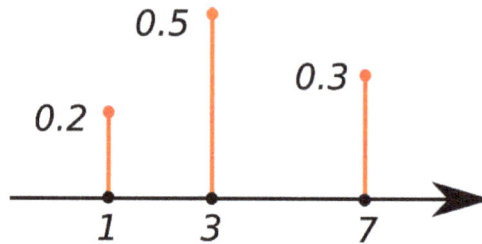

The probability mass function of a discrete probability distribution. The probabilities of the singletons {1}, {3}, and {7} are respectively 0.2, 0.5, 0.3. A set not containing any of these points has probability zero.

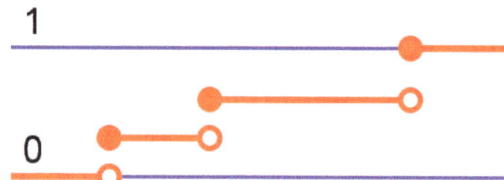

The cdf of a discrete probability distribution, ...

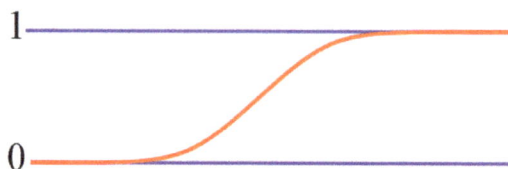

... of a continuous probability distribution, ...

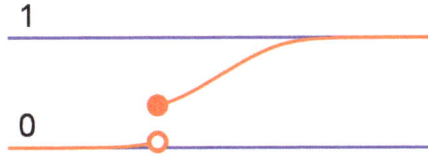

... of a distribution which has both a continuous part and a discrete part.

A discrete probability distribution is a probability distribution characterized by a probability mass function. Thus, the distribution of a random variable X is discrete, and X is called a discrete random variable, if

$$\sum_u \Pr(X = u) = 1$$

as u runs through the set of all possible values of X. A discrete random variable can assume only a finite or countably infinite number of values. For the number of potential values to be countably infinite, even though their probabilities sum to 1, the probabilities have to decline to zero fast enough. For example, if $\Pr(X = n) = \frac{1}{2^n}$ for n = 1, 2, ..., we have the sum of probabilities $1/2 + 1/4 + 1/8 + ... = 1$.

Well-known discrete probability distributions used in statistical modeling include the Poisson distribution, the Bernoulli distribution, the binomial distribution, the geometric distribution, and the negative binomial distribution. Additionally, the discrete uniform distribution is commonly used in computer programs that make equal-probability random selections between a number of choices.

Measure Theoretic Formulation

A measurable function $X : A \to B$ between a probability space $(A, \mathcal{A}, P)$ and a measurable space $(B, \mathcal{B})$ is called a discrete random variable provided its image is a countable set and the pre-image of singleton sets are measurable, i.e., $X^{-1}(b) \in \mathcal{A}$ for all $b \in B$. The latter requirement induces a probability mass function $f_X : X(A) \to \mathbb{R}$ via $f_X(b) := P(X^{-1}(b))$. Since the pre-images of disjoint sets are disjoint

$$\sum_{b \in X(A)} f_X(b) = \sum_{b \in X(A)} P(X^{-1}(b)) = P\left(\bigcup_{b \in X(A)} X^{-1}(b) \right) = P(A) = 1.$$

This recovers the definition given above.

Cumulative Density

Equivalently to the above, a discrete random variable can be defined as a random variable whose cumulative distribution function (cdf) increases only by jump discontinuities—that is, its cdf increases only where it "jumps" to a higher value, and is constant between those jumps. The points where jumps occur are precisely the values which the random variable may take.

Delta-function Representation

Consequently, a discrete probability distribution is often represented as a generalized probability density function involving Dirac delta functions, which substantially unifies the treatment of continuous and discrete distributions. This is especially useful when dealing with probability distributions involving both a continuous and a discrete part.

Indicator-function Representation

For a discrete random variable X, let u_0, u_1, ... be the values it can take with non-zero probability. Denote

$$\Omega_i = X^{-1}(u_i) = \{\omega : X(\omega) = u_i\}, i = 0, 1, 2, \ldots$$

These are disjoint sets, and by formula (1)

$$\Pr\left(\bigcup_i \Omega_i\right) = \sum_i \Pr(\Omega_i) = \sum_i \Pr(X = u_i) = 1.$$

It follows that the probability that X takes any value except for u_0, u_1, ... is zero, and thus one can write X as

$$X(\omega) = \sum_i u_i 1_{\Omega_i}(\omega)$$

except on a set of probability zero, where 1_A is the indicator function of A. This may serve as an alternative definition of discrete random variables.

Continuous Probability Distribution

A continuous probability distribution is a *probability distribution* that has a cumulative distribution function that is continuous. Most often they are generated by having a probability density function. Mathematicians call distributions with probability density functions absolutely continuous, since their cumulative distribution function is absolutely continuous with respect to the Lebesgue measure λ. If the distribution of X is continuous, then X is called a continuous random variable. There are many examples of continuous probability distributions: normal, uniform, chi-squared, and others.

Intuitively, a continuous random variable is the one which can take a continuous range of values—as opposed to a discrete distribution, where the set of possible values for the random variable is at most countable. While for a discrete distribution an event with probability zero is impossible (e.g., rolling $3\frac{1}{2}$ on a standard dice is impossible, and has probability zero), this is not so in the case of a continuous random variable. For example, if one measures the width of an oak leaf, the result of 3½ cm is possible; however, it has probability zero because uncountably many other potential values exist even between 3 cm and 4 cm. Each of these individual outcomes has probability zero, yet the probability that the outcome will fall into the interval (3 cm, 4 cm) is nonzero. This apparent paradox is resolved by the fact that the probability that X attains some value within an infinite set, such as an interval, cannot be found by naively adding the probabilities for individual values. Formally, each value has an infinitesimally small probability, which statistically is equivalent to zero.

Formally, if X is a continuous random variable, then it has a probability density function $f(x)$, and therefore its probability of falling into a given interval, say $[a, b]$ is given by the integral

$$\Pr[a \le X \le b] = \int_a^b f(x)dx$$

In particular, the probability for X to take any single value a (that is $a \le X \le a$) is zero, because an integral with coinciding upper and lower limits is always equal to zero.

The definition states that a continuous probability distribution must possess a density, or equivalently, its cumulative distribution function be absolutely continuous. This requirement is stronger than simple continuity of the cumulative distribution function, and there is a special class of distributions, singular distributions, which are neither continuous nor discrete nor a mixture of those. An example is given by the Cantor distribution. Such singular distributions however are never encountered in practice.

Note on terminology: some authors use the term "continuous distribution" to denote the distribution with continuous cumulative distribution function. Thus, their definition includes both the (absolutely) continuous and singular distributions.

By one convention, a probability distribution μ is called *continuous* if its cumulative distribution function $F(x) = \mu(-\infty, x]$ is continuous and, therefore, the probability measure of singletons $\mu\{x\}=0$ for all x.

Another convention reserves the term *continuous probability distribution* for absolutely continuous distributions. These distributions can be characterized by a probability density function: a non-negative Lebesgue integrable function f defined on the real numbers such that

$$F(x) = \mu(-\infty, x] = \int_{-\infty}^x f(t)dt.$$

Discrete distributions and some continuous distributions (like the Cantor distribution) do not admit such a density.

Some properties

- The probability distribution of the sum of two independent random variables is the convolution of each of their distributions.

- Probability distributions are not a vector space—they are not closed under linear combinations, as these do not preserve non-negativity or total integral 1— but they are closed under convex combination, thus forming a convex subset of the space of functions (or measures).

Kolmogorov Definition

In the measure-theoretic formalization of probability theory, a random variable is defined as a measurable function X from a probability space $(\Omega, \mathcal{F}, P)$ to measurable space $(\mathcal{X}, \mathcal{A})$. A probability distribution of X is the pushforward measure X_*P of X, which is a probability measure on $(\mathcal{X}, \mathcal{A})$ satisfying $X_*P = PX^{-1}$.

Random Number Generation

A frequent problem in statistical simulations (the Monte Carlo method) is the generation of pseudo-random numbers that are distributed in a given way. Most algorithms are based on a pseudorandom number generator that produces numbers X that are uniformly distributed in the interval [0,1]. These random variates X are then transformed via some algorithm to create a new random variate having the required probability distribution.

Applications

The concept of the probability distribution and the random variables which they describe underlies the mathematical discipline of probability theory, and the science of statistics. There is spread or variability in almost any value that can be measured in a population (e.g. height of people, durability of a metal, sales growth, traffic flow, etc.); almost all measurements are made with some intrinsic error; in physics many processes are described probabilistically, from the kinetic properties of gases to the quantum mechanical description of fundamental particles. For these and many other reasons, simple numbers are often inadequate for describing a quantity, while probability distributions are often more appropriate.

As a more specific example of an application, the cache language models and other statistical language models used in natural language processing to assign probabilities to the occurrence of particular words and word sequences do so by means of probability distributions.

Common Probability Distributions

The following is a list of some of the most common probability distributions, grouped by the type of process that they are related to.

All of the univariate distributions below are singly peaked; that is, it is assumed that the values cluster around a single point. In practice, actually observed quantities may cluster around multiple values. Such quantities can be modeled using a mixture distribution.

Related to Real-valued Quantities that Grow Linearly (e.g. Errors, Offsets)

- Normal distribution (Gaussian distribution), for a single such quantity; the most common continuous distribution

Related to Positive Real-valued Quantities that Grow Exponentially (e.g. Prices, Incomes, Populations)

- Log-normal distribution, for a single such quantity whose log is normally distributed

- Pareto distribution, for a single such quantity whose log is exponentially distributed; the prototypical power law distribution

Related to Real-valued Quantities that are Assumed to be Uniformly Distributed Over a (Possibly Unknown) Region

- Discrete uniform distribution, for a finite set of values (e.g. the outcome of a fair die)

- Continuous uniform distribution, for continuously distributed values

Related to Bernoulli Trials (yes/no Events, with a Given Probability)

- Basic distributions:
  - Bernoulli distribution, for the outcome of a single Bernoulli trial (e.g. success/failure, yes/no)

  - Binomial distribution, for the number of "positive occurrences" (e.g. successes, yes votes, etc.) given a fixed total number of independent occurrences

  - Negative binomial distribution, for binomial-type observations but where the quantity of interest is the number of failures before a given number of successes occurs

- o Geometric distribution, for binomial-type observations but where the quantity of interest is the number of failures before the first success; a special case of the negative binomial distribution

- Related to sampling schemes over a finite population:

  - o Hypergeometric distribution, for the number of "positive occurrences" (e.g. successes, yes votes, etc.) given a fixed number of total occurrences, using sampling without replacement

  - o Beta-binomial distribution, for the number of "positive occurrences" (e.g. successes, yes votes, etc.) given a fixed number of total occurrences, sampling using a Polya urn scheme (in some sense, the "opposite" of sampling without replacement)

Related to Categorical Outcomes (Events with K Possible Outcomes, with a Given Probability for each Outcome)

- Categorical distribution, for a single categorical outcome (e.g. yes/no/maybe in a survey); a generalization of the Bernoulli distribution

- Multinomial distribution, for the number of each type of categorical outcome, given a fixed number of total outcomes; a generalization of the binomial distribution

- Multivariate hypergeometric distribution, similar to the multinomial distribution, but using sampling without replacement; a generalization of the hypergeometric distribution

Related to Events in a Poisson Process (Events that Occur Independently with a given Rate)

- Poisson distribution, for the number of occurrences of a Poisson-type event in a given period of time

- Exponential distribution, for the time before the next Poisson-type event occurs

- Gamma distribution, for the time before the next k Poisson-type events occur

Related to the Absolute Values of Vectors with Normally Distributed Components

- Rayleigh distribution, for the distribution of vector magnitudes with Gaussian distributed orthogonal components. Rayleigh distributions are found in RF signals with Gaussian real and imaginary components.

- Rice distribution, a generalization of the Rayleigh distributions for where there is a stationary background signal component. Found in Rician fading of radio signals due to multipath propagation and in MR images with noise corruption on non-zero NMR signals.

Related to Normally Distributed Quantities Operated with Sum of Squares (for Hypothesis Testing)

- Chi-squared distribution, the distribution of a sum of squared standard normal variables; useful e.g. for inference regarding the sample variance of normally distributed samples

- Student's t distribution, the distribution of the ratio of a standard normal variable and the square root of a scaled chi squared variable; useful for inference regarding the mean of normally distributed samples with unknown variance

- F-distribution, the distribution of the ratio of two scaled chi squared variables; useful e.g. for inferences that involve comparing variances or involving R-squared (the squared correlation coefficient)

Useful as Conjugate Prior Distributions in Bayesian Inference

- Beta distribution, for a single probability (real number between 0 and 1); conjugate to the Bernoulli distribution and binomial distribution

- Gamma distribution, for a non-negative scaling parameter; conjugate to the rate parameter of a Poisson distribution or exponential distribution, the precision (inverse variance) of a normal distribution, etc.

- Dirichlet distribution, for a vector of probabilities that must sum to 1; conjugate to the categorical distribution and multinomial distribution; generalization of the beta distribution

- Wishart distribution, for a symmetric non-negative definite matrix; conjugate to the inverse of the covariance matrix of a multivariate normal distribution; generalization of the gamma distribution

Distribution

Probability Distributions

In statistics, a probability distribution gives either the probability of each value of a random variable (when the variable is discrete), or the probability of the value falling within a particular interval (when the variable is continuous). The probability distribution describes the range of possible values that a random variable can attain and the probability that the value of the random variable is within any (measurable) subset of that range.

A probability distribution gives important information about the data, how the values are changing, whether they are bunched together or spread out, and whether they are symmetrically disposed on the X-axis or not. Distribution also tells the relative frequency or proportion of various X values in the population in the same way that a histogram gives information about a sample. We now describe the distributions that are commonly used in addressing water resources problems.

Commonly used distributions in hydrology are the Normal, Log Normal, Extreme Value type-1 (Gumbel or EV1), Gamma, Pearson Type - III, and Log Pearson Type - III distributions. A brief description of these distributions is given below.

Normal Distribution

It is also known as the Gaussian distribution. When a hydrologic variable, integrated over a large time period is used in analysis, the variable is expected to follow a normal distribution. The normal distribution has a symmetrical bell-shaped probability density function. The two parameters of the normal distribution are mean μ and standard deviation σ. Its PDF can be expressed as

$$f(x) = \frac{1}{\sigma\sqrt{2\pi}} \exp\left[-\frac{(x-\mu)^2}{2\sigma^2}\right] \quad -\infty < x < \infty$$

the cumulative density function (CDF) of the normal distribution is:

$$F(x) = \frac{1}{\sigma\sqrt{2\pi}} \int_{-\infty}^{x} \exp\left[-\frac{(x-\mu)^2}{2\sigma^2}\right] du$$

The origin of the normal organization lies in the central limit theorem which states that if a sequence of random variables $x_i, i = 1, 2......, n$ are independently and identically distributed with mean μ and standard deviation σ then the distribution of n such random variables $Y = \sum_{i=1}^{n} x$ tends to the normal distribution with mean $n\mu$ and standard deviation $\sqrt{n}\sigma$ as n becomes large. This theorem holds good irrespective of the probability distribution of x.

The reduced variate of the normal distribution is defined as $Z = (x - \mu)/\sigma$. The properties of the reduced variate are mean = 0, standard deviation $\sigma_z = 1$, and coefficient of skewness = 0. Figure shows the normal distribution and the area under the curve for three values of the reduced variate. As shown, the area under the curve within $\mu \pm \sigma$ is 68.27%, within $\mu \pm 2\sigma$ is 95.45 and it is 99.73 within $\mu \pm 3\sigma$.

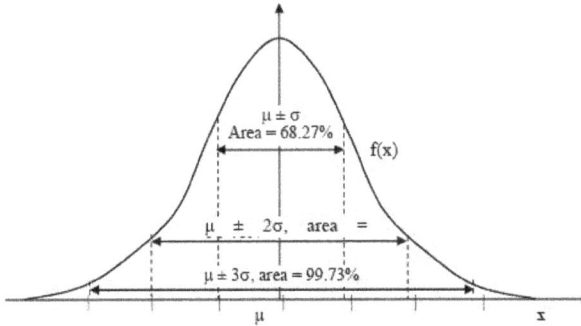

The normal distribution and the area for three values of the standard variate.

Among the probability distributions used in hydrology, the normal distribution is the most widely. It is also employed in the analysis of variance, estimation of random errors of hydrologic measurements, hypothesis testing, synthetic generation of random numbers, etc. A random variable that is made up of the sum of many small independent effects is expected to follow a normal distribution. Many hydrologic variables are not normally distributed, but transformations can, in many cases, make them approximately normally distributed. When there is increase in the time interval over which a hydrologic variable is measured, the variable approximately follows a normal distribution because the number of causative effects increases.

Example: Assuming that the data of Sabarmati River follows the normal distribution, find the parameters of the distribution and plot it.

Solution: For the data of Sabarmati river, the mean and SD are:

Mean of the data $\bar{x}$ = 665.37 million cubic m.
Standard deviation σ = 346.9 million cubic m.
Coefficient of variation C_V = 346.9/665.37 = 0.521.
Coefficient of skewness C_s = 0.76 (positively skewed).
Kurtosis C_k = 3.65.

Figure shows the plot of the probability distribution of Sabarmati River data.

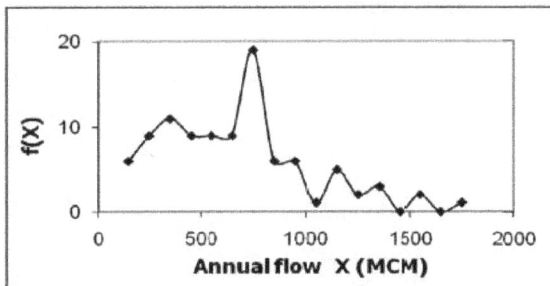

Probability distribution of Sabarmati River data.

For the Sabarmati data, the coefficient of skewness C_s is 0.76 or the data is positively skewed. This is easily verified by figure. Further, kurtosis C_k for the data is 3.65 (kurtosis for the normal distribution is 3). Again, this can also be verified from figure.

Log-Normal Distribution

The log-normal distribution is the probability distribution of a random variable whose logarithm is normally distributed. Let X be a random variable with a normal distribution, then $Y = \exp(X)$ has a log-normal distribution. In other words, if Y is log-normally distributed, then $X = \log(Y)$ is normally distributed. When a random variable represents a process that is the resultant of multiplicative product of many small effects each of which is positive, then it can be expressed the sum of logarithms of these small effects. The logarithm of such a random variable can be expected to follow a normal distribution. Hence, if the variable is transformed to the log domain, it is likely to follow the normal distribution. An advantage of the log-normal distribution is that it is often useful to represent quantities that cannot have negative values. It has proven useful to model rainfall amounts, size distributions of aerosol particles, etc.

The PDF of the log-normal distribution is

$$f(x) = \frac{1}{x\sigma_y \sqrt{2\pi}} \exp\left[-\frac{\left(\ln x - \mu_y\right)^2}{2\sigma_y^2} \right] \quad x > 0$$

The log-normal distribution has two parameters μ_y and σ_y which can be estimated by transforming all x_i's to y_i's by

$$y_i = \ln x_i$$

Extreme Value Type 1 (EV1) Distribution

Let a series of large number of (N) observations of random variable be subdivided into n subsamples of size m each, such that $N = nm$. Each subseries shall have two extreme values: one maximum and one minimum corresponding to, for example, floods and droughts. Gumbel (1958) showed that the n largest values of subsamples asymptotically follow an extreme value type 1 (EV1) distribution. This distribution, also known as the Gumbel distribution or double negative exponential distribution, is widely used for frequency analysis of floods, maximum rainfall, etc. This distribution is essentially a log-normal distribution with constant skewness (approximately 1.14). Its PDF and CDF are as follows:

$$f(x) = \alpha \exp\left\{ -\alpha(x - \beta) - \exp\left[-\alpha(x - \beta) \right] \right\} \quad -\infty < x < \infty; \quad -\infty < \beta < \infty; \quad \alpha > 0$$

$$F(x) = \exp\{-\exp[-\alpha(x-\beta)]\}$$

where α, and β are scale and location parameters. The estimates of parameters using the method of moments are:

$$\hat{\alpha} = \frac{1.283}{s}; \quad \hat{\beta} = \bar{X} - 0.45s$$

It has been shown that the EVI distribution is a special case of a distribution known as the Generalized Extreme Value (GEV) distribution. The CDF of the GEV distribution is given by

$$F(x) = \exp\left[-\left(1 - k\frac{x-u}{\alpha}\right)^{1/k}\right]$$

Where k, u, and α are the parameters of the distribution. When k=0, we get the EV1 distribution. For k<0, the distribution known as EV2 and it is known as EV3 distribution when k>0.

According to Gumbel, the probability that an event with magnitude larger than x0 occurs is (Subramanya 2008):

$$\text{Prob } (X \geq x_o) = 1 - \exp[-\exp(-y)]$$

where y is the reduced variate, given as

$$y = \alpha(x - \beta),$$
$$\beta = \bar{x} - 0.45\sigma_x$$
$$\alpha = 1.2825 / \sigma_x$$

Substituting the values of a and α

$$y = 1.285(x - \bar{x}) / \sigma_x + 0.577$$

The expression for the reduced variate y for return period T is

$$y = -[0.834 + 2.303 \log(\log\frac{T}{T-1})]$$

Now we can compute variate x with return period T by

$$X_T = \bar{x} + k\sigma_x \tag{a}$$

Where $k = (y_T - 0.577)/1.2825$ (b)

Equations (a) and (b) assume that a large data series are available to compute the various statistics. However in practice, the record length is finite. In such cases, the following equation may be used:

$$X_T = \bar{x} + k_n \sigma_{n-1}$$

where σ_{n-1} is the standard deviation of the sample of size n. Frequency factor for use with sample of size n is given as

$$k_n = (y_T - \bar{y}_n)/s_n$$

Where $\bar{y}_n$ and s_n are reduced mean and reduced standard deviations which are functions of n. Values of these can be obtained from standard tables that are widely available. Note that as $n \to \infty, \bar{y}_n \to 0.577$ and $s_n \to 1.2825$.

Example: From the flow data of a river, the mean and standard deviation were computed and these turned out to be 660 million cubic m and 330 million cubic m, respectively. Find the value of parameters of EV1 distribution.

Solution: The mean and standard deviation of the data are 665.37 million cubic m and 346.9 million cubic m, respectively. Therefore, the estimates by the method of moment are:

$$\alpha = 1.2825/330 = 0.0039$$
$$\text{and} \quad \beta = 660 - 0.45 * 330 = 511.5$$

Example: Annual maximum flood discharge data of a river was available for 30 years. Mean and standard deviation were 5250 m³/s and 1650 m³/s. Compute the flood discharge with a return period of 100 years by using the Gumbel Extreme Value 1 distribution.

Solution: From standard tables, for n=30 years

$$y_n = 0.5362, s_n = 1.1124.$$

Hence
$$y_T = -[0.834 + 2.303 \log(\log \frac{100}{99})] = 4.601$$

$$k_{100} = (4.601 - 0.5362)/1.1124 = 3.654$$

$$X_{100} = 5250 + 3.654 * 1650 = 11279 \text{m}^3/\text{s}$$

Log Pearson Type - III (LP3) Distribution

Log Pearson Type III distribution was found to give good results in numerous studies dealing with flood peak data. This distribution is the standard distribution for flood frequency analysis in the USA since its use for flood frequency analysis was recommended by the US Water Resources Council.

LP3 is a three-parameter distribution and is widely used in hydrology. Its parameters are related to mean, standard deviation, and skewness.

$$f(x) = \frac{1}{a\Gamma(b)}\left(\frac{x-c}{a}\right)^{b-1}\exp\left(-\frac{x-c}{a}\right) \tag{i}$$

wherea, b, and c are scale, shape, and location parameters, respectively, and $\Gamma(b)$ is a gamma function. If c=0, this distribution becomes a two-parameter gamma distribution. Parameters a, b, and c are related to mean, standard deviation, and coefficient of skewness as (method of moment estimates)

$$a = \sigma / \sqrt{b}$$
$$b = \left(2 / C_s\right)^2$$
$$c = \mu - \sigma\sqrt{b}$$

To determine flood for a return period T by using the LP3 distribution, the procedure described below is followed.

First of all, the frequency factor, KT is computed by (Chow et al. 1988):

$$K_T = z + (z^2 - 1)k + (z^3 - 6z)k^2/3 + (z^2 - 1)k^3 + zk^4 + k^5/3$$

Where $k = C_s/6$. To complete z for a given return period T, exceedance probability p is obtained as p = 1/T. Now, complete a variable w as

$$w = \sqrt{\ln\left(1/P^2\right)} \qquad 0 < P \le 0.5$$

Now z is calculated by (Abramowitz and Stegun, 1965)

$$z = w - \frac{2.515517 + 0.802853w + 0.010328w^2}{1 + 1.432788w + 0.189269w^2 + 0.001308w^3} \tag{ii}$$

when p>0.5, p in eq. (ii) is replaced by (1-p) and the negative sign is put before z computed by eq. (iii). Now, by following the frequency factor method, the flood for the return period T years is computed by:

$$y_T = \bar{y} + K_T s_y \tag{iii}$$

Example: For the data of example, find the parameters of the Pearson Type III distribution.

Solution: The estimates of parameters using the method of moments are

$$b = (2 / C_s)^2 = (2 / 0.76)^2 = 6.93 \text{ million cubic m.}$$

$$a = 346.9\sqrt{6.93} = 131.78 \text{ million cubic m.}$$

$$c = 665.37 - 346.9 * \sqrt{6.93} = -247.84$$

Example: Logarithms of the annual flood peak data of a river were taken and the mean was 4.146, SD was 0.403 and C_s = -0.07. Find 50 year return period flood by using the LP3 distribution.

Solution First we find the value of K_{50} by the following equation:

$$K_{50} = z + (z^2 - 1)k + (z^3 - 6z)k^2 / 3 + (z^2 - 1)k^3 + zk^4 + k^5 / 3$$

Here $k = C_s / 6 = -0.07 / 6 = -0.0117$. For $T = 50, p = 1 / 50 = 0.02$.

$$w = \sqrt{\ln(1 / 0.02^2)} = 2.797$$

From eq. (ii)

$$z = 2.797 - \frac{2.515517 + 0.802853 * 2.797 + 0.010328 * 2.797^2}{1 + 1.432788 * 2.797 + 0.189269 * 2.797^2 + 0.001308 * 2.797^3} = 2.054$$

Now K_{50} is calculated as

$$K_{50} = 2.054 + \left(2.054^2 - 1\right) * \left(-0.0117\right) + \left(2.054^3 - 6 * 2.054\right) * \left(-0.0117\right)^2 / 3$$

$$+ \left(2.054^2 - 1\right) * \left(-0.0117\right)^3 + 2.054 * \left(-0.0117\right)^4 + \left(-0.0117\right)^5 / 3$$

$$= 2.016$$

Hence, $y_{50} = 4.146 + 2.016 * 0.403 = 4.959$

So, the 50-year flood $x_{50} = (10)^{4.959} = 90942$.

Discrete Probability Distributions

The use of discrete probability distributions is restricted generally to those random events in which the outcome can be described as success or failure, i.e., there are only two mutually exclusive events in an experiment. Moreover, the successive trials are independent and the probability of success remains constant from trial to trial. The binomial or Poisson distributions can be used to find the probability of occurrence of an event r times in n successive years.

Binomial Distribution

This distribution arises in Bernoulli processes where in any trial; the event may or may not take place. The probability of occurrence of the event is the same from one trial to another. This distribution usually occurs while dealing with complementary events. A common example is tossing of coins in which the probability of head appearing is the same in each trial. The occurrence of wet and dry days over a given time interval is also a complementary event. The probability of occurrence of the event r times in n successive years is given by:

$$P_{r,n} = {}^nC_r P^r q^{n-r} = \frac{n!}{r!(n-r)!} p^r q^{n-r}$$

where $P_{r,n}$ is the probability of a random event of a given magnitude and exceedance probability P occurring r times in n successive years. The probability of the event not occurring at all in n successive years is:

$$P_{o,n} = q^n = (1-p)^n$$

The probability of an event occurring at least once in n successive years:

$$P_1 = 1 - q^n = 1 - (1-p)^n$$

Example: An analysis of data on the maximum one-day rainfall depth at a station indicated that a depth of 280 mm had a return period of 50 years. Determine the probability of a one-day rainfall depth equal to or greater than 280 mm occurring (a) once in 20 successive years, and (b) two times in 15 successive years.

Solution: Here, $P = 1/50 = 0.02$.

a) In the first case, $n = 20, r = 1$. Therefore, from equation

$$P_{1,20} = \frac{20!}{19!1!} * (0.02) * (0.98)^{19} = 0.272.$$

b) In this case, $n = 15, r = 2$. Therefore,

$$P_{2,15} = \frac{15!}{13!2!} * (0.02^2) * (0.980)^{13} = 0.0292.$$

Example: What is the probability that a 5-year flood will not occur at all in a 10-year period?

Solution: Here, $p = 1/5 = 0.2, n = 10,$ and $r = 0$. Hence the probability is

$$P_{0,10} = \frac{10!}{0!10!} * 0.2^0 * (0.8)^{10} = 0.1074$$

Poisson Distribution

The Poisson distribution is a limiting form of the binomial distribution when p is very small and n is very large, and np tends to a constant value λ. This may happen when the interval over which the Bernoulli process is defined gets smaller and smaller and the number of trials becomes greater and greater, keeping np constant. The Poisson distribution has only one parameter λ that denotes the expected mean frequency of occurrence of some event in a given time t. The probability distribution of the number of events in a given time is

$$P(X = x) = \frac{\lambda^x \exp(-\lambda)}{x!}, \lambda > 0, \ x = 0,1,2...$$

The CDF of the Poisson distribution is

$$P(X \leq x) = \sum_{i=0}^{x} \frac{\lambda^i \exp(-\lambda)}{i!}$$

The conditions for application of Poisson distribution are: a) the number of events is discrete, b) two events cannot coincide, c) the mean number of events per unit time is constant, and d) events are independent. Thus, it can be applied to following situations with p relatively small and n relatively large to determine the probability of:

(i)   Droughts in a given time period,

(ii)   Number of rainy days at a given location,

(iii) Probability of rare flood events, and

(iv)   Probability of reservoir being empty in any one year out of a long period of record.

Parameter Estimation Methods

A number of methods have been developed to estimate parameters of hydrologic models. Some commonly used methods in hydrology include: (1) method of moments; (2) method of probability weighted moments; (4) L-moments; (5) maximum likelihood estimation; and (6) least squares method. Each of these methods is discussed here.

Method of Moments

This method is very commonly employed to estimate parameters of linear hydrologic models. This method is based on the premise that when the parameters of a probability distribution are estimated correctly, the moments of the probability density function are equal to the corresponding moments of a sample data. Nash (1959) developed the theorem of moments which relates the moments of input, output and impulse response functions of linear hydrologic models.

Let X be a continuous variable and $f(x)$ its function satisfying some necessary conditions. The rth moment of $f(x)$ about an arbitrary point 'a' is denoted as $M_r^a(f)$. The rth moment of the function $f(x)$ can be expressed as

$$M_r^a(f) = \int_{-\infty}^{\infty} (x-a)^r f(x)dx$$

Figure shows the definition of various terms used in the above equation.

Consider the special case when r = 0. In this case, the above eq. gives

$$M_0^a = \int_{-\infty}^{\infty} (x-a)^0 f(x)dx = \int_{-\infty}^{\infty} f(x)dx = 1$$

Thus, the zero-order moment is the area under the curve defined by $f(x)$ subject to $-\infty < x < \infty$. For probability distribution, this area is unity. If $r = 1$, then

$$M_1^a = \int_{-\infty}^{\infty} (x-a)^1 f(x)dx = \mu - a$$

where μ is the mean. If the moment is taken around the origin, then $a = 0$, and the first moment gives the mean. When $a = \mu$, the rth moment about the mean is expressed by

$$M_r^\mu = \int_{-\infty}^{\infty} (x-\mu)^r f(x)dx$$

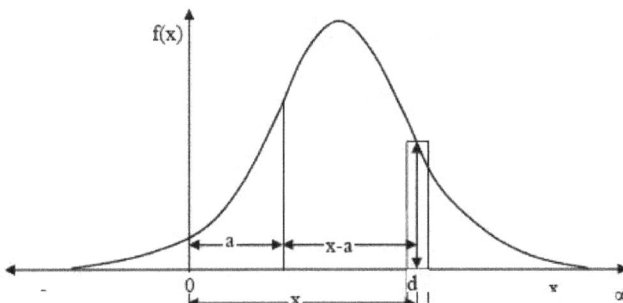

Concept of moment of a function f(x) about an arbitrary point.

For simplicity of notation, we drop the superscript if the moment is taken about the origin o and the familiar terminology of the moments can be written as follows:

M_0 =Area

M_1 = Mean

M_2^μ = Variance,

M_3^μ = Measurement of skewness of the function

M_4^μ = Kurtosis,

Method of Moments for Discrete Systems

For a discrete function, represented as $f_j, j = -\infty, ..., -1, 0, 1, ..., \infty$, the rth moment about any arbitrary point can be defined in an analogous manner as for continuous functions. The rth moment about the origin, is defined as

$$M_r = \sum_{m=-\infty}^{\infty} m^r f_m$$

Figure depicts the concept of moment of a discrete function.

Example: The frequency table of annual flows of Sabarmati River is given table. Find the mean and variance of the data by using the method of moments.

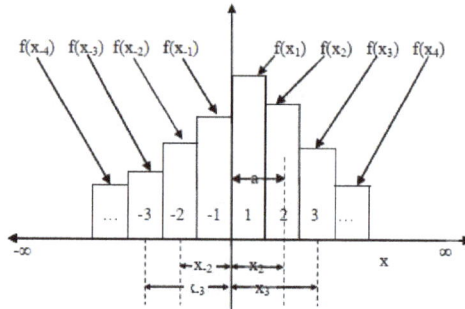

Concept of moment of a discrete function about an arbitrary point 'a'.

Table: Frequency table of annual flows of Sabarmati River.

Discharge range	Frequency	Discharge range	Frequency
100-200	6	200-300	9
300-400	11	400-500	9
500-600	9	600-700	9
700-800	19	800-900	6
900-1000	6	1000-1100	1
1100-1200	5	1200-1300	2
1300-1400	3	1400-1500	0
1500-1600	2	1600-1700	0
1700-1800	1		

Solution: The sum of all frequencies = (6+11+9+19+ ...) = 98. For the first range of discharge (100-200), the mean value is 150; for the second range (200-300), it is 250 and so on.

Hence, the first moment of the data = (150*6 + 250*9 + 350*11 + ...+ 1750*1)/98

$$= 664.29 \text{ cumec.}$$

This is the mean (x_m) of the annual flows.

The second moment about the mean gives the variance of the data.

$$\text{Second } M_2^{\mu} = \sum_{i=1}^{N} \frac{(x_i - x_m)^2 f(x_i)}{N} \text{ moment}$$

$$= [(150\text{-}664)^2*6 + (250 - 664)^2*9 + ... + (1750\text{-}664)^2*1]/98$$

$$= 120000 \text{ cumec}^2.$$

Hence, the standard deviation (s) = $(120000)^{0.5}$ =346.61 cumec.

This leads to the coefficient of variation c_v = s/x_m =346.61/664.29 = 0.52.

Method of Maximum Likelihood (MLE)

The maximum likelihood (ML) estimation method is widely accepted as one of the most powerful parameter estimation methods. Asymptotically, the ML parameter estimates are unbiased, have minimum variance, and are normally distributed, while in some cases these properties hold for small samples. The MLE method has been extensively used for estimating parameters of frequency distributions as well as fitting conceptual models.

Let $f(x; a_1, a_2,... a_m)$ be a PDF of the random variable X with parameters a_i, i=1, 2, ..., m, to be estimated. For a random sample of data $x_1, x_2, ...x_n$, drawn from this probability density function, the joint PDF is defined as

$$f(x_i, x_2,....x_n; a_1, a_2,....a_m) = \prod_{i=1}^{n} f(x_i; a_1, a_2,.....a_m)$$

Interpreted conceptually, the probability of obtaining a given value of X, say x_1, is proportional to f(x; a_1, a_2,... a_m). Likewise, the probability of obtaining the random sample x_1, x_2,... x_n from the population of X is proportional to the product of the individual probability densities or the joint PDF. This joint PDF is called the likelihood function, denoted by L.

$$L = \prod_{i=1}^{n} f(x_1; a_1, a_2,.....a_m)$$

where the parameters a_i, i=1,2,...m, are unknown.

By maximizing the likelihood that the sample under consideration is the one that would be obtained if n random observations were selected from f(x; a$_1$, a$_2$, ... a$_m$), the unknown parameters are determined, and hence the name of the method. The values of parameters so obtained are known as MLE estimators. Since the logarithm of L attains its maximum for the same values of ai, i = 1, 2, ... m, as does L, the MLE function can also be expressed as

$$\ln L = L^* = \ln \prod_{i=1}^{n} f(x_1; a_1, a_2, ... a_m) = \sum_{i=1}^{n} \ln f(x_1; a_1, a_2, ... a_m)$$

Frequently ln[L] is maximized, for it is many times easier to find the maximum of the logarithm of the maximum likelihood function than that of the normal L.

The procedure for estimating parameters or determining the point where the MLE function achieves its maximum involves differentiating L or $ln\ L$ partially with respect to each parameter and equating each differential to zero. This results in as many equations as the number of unknown parameters. For m unknown parameters, we get

$$\frac{\partial L(a_1, a_2, ... a_m)}{\partial a_1} = 0$$

$$\frac{\partial L(a_1, a_2, ... a_m)}{\partial a_2} = 0$$
$$\frac{\partial L(a_1, a_2, ... a_m)}{\partial a_m} = 0$$

These m equations in m unknowns are then solved for the m unknown parameters.

Example: Using the method of maximum likelihood, find the parameter α of the exponential distribution for the data of the Sabarmati River in India, given in example.

Solution: The probability density function of the one-parameter exponential distribution is given by

$$f_X(x) = \alpha exp(-\alpha x)$$

The likelihood function is given by

$$L(\alpha) = \prod_{i=1}^{n} \alpha \exp(-\alpha x_i) = \alpha^n \exp\left(-\alpha \sum_{i=1}^{n} x_i\right)$$

This can be used to form the log-likelihood function:

$$\ln L(\alpha) = n \ln(\alpha) - \alpha \left(\sum_{i=1}^{n} x_i\right)$$

where n is the sample size. Differentiating above equation with respect to α:

$$\frac{d \ln L(\alpha)}{d\alpha} = \frac{n}{\alpha} - \sum_{i=1}^{n} x_i = 0$$

This yields

$$\alpha = n \Big/ \left(\sum_{i=1}^{n} x_i \right) = \frac{1}{\bar{x}}$$

In example, the mean of the data was found to be 664.29 cumec. Hence, the estimate of α is:

$$\alpha = 1/664.29 = 1.51 \times 10^{-3} \text{ cumec}^{-1}.$$

Method of Least Squares

The method of least squares (MOLS) is one of the most frequently used parameter estimation methods in hydrology. Natale and Todini (1974) presented constrained MOLS for linear models in hydrology.

Let there be a function Y = f(X; a_1, a_2,... a_m), where a_i, i = 1, 2, ... m, are parameters to be estimated. The method of least squares (MOLS) involves estimating parameters by minimizing the sum of squares of all deviations between observed and computed values of Y. Mathematically, this sum D can be expressed as

$$D = \sum_{i=1}^{n} d_i^2 = \sum_{i=1}^{n} [y_0(i) - y_c(i)]^2 = \sum_{i=1}^{n} [y_0(i) - f(x; a_1, a_2, ... a_m)]^2$$

where $y_0(i)$ is the ith observed value of Y, $y_c(i)$ is the ith computed value of Y, and n > m is the number of observations. The minimum of D in equation can be obtained by differentiating D partially with respect to each parameter and equating each differential to zero, e.g.,

$$\frac{\partial \sum_{i=1}^{n} [y_0(i) - f(x_i; a_1, a_2, ... a_m)]^2}{\partial a_1} = 0$$

The resulting m equations, usually called the normal equations, are then solved for estimation of m parameters. This method is frequently used to estimate parameters of linear regression model.

Method of L-Moments

Greenwood et al. (1979) introduced the method of probability weighted moments (PWM) and showed its usefulness in deriving explicit expressions for parameters of distributions whose inverse forms $X = X(F)$ can be explicitly defined. They derived relations between parameters and PWMs for Generalized Lambda, Wakeby, Weibull, Gumbel, Logistic and Kappa distributions. However, the probability-weighted moments characterize a distribution but are not meaningful by themselves.

L-moments were developed by Hosking (1986) as functions of PWMs which provide a descriptive summary of the location, scale, and shape of the probability distribution. These moments are analogous to ordinary moments and are expressed as *linear* combinations of order statistics, hence the name. They can also be expressed by linear combinations of probability- weighted moments. Thus, the ordinary moments, the probability weighted moments, and L- moments are related to each other. L-moments are known to have several important advantages over ordinary moments. L-moments have less bias than ordinary moments because they are linear combinations of ranked observations. As an example, variance (second moment) and skewness (third moment) involve squaring and cubing of observations, respectively, which compel them to give greater weight to the observations far from the mean. As a result, they result in substantial bias and variance.

The first L-moment denoted as λ_1 is the arithmetic mean:

$$\lambda_1 = E[X]$$

Let us consider a sample of size n and arrange the data such that $X_{(i|n)}$ is the ith largest observation; clearly i = n will be the largest value. Then, for any distribution, the second L- moment, λ_2, is a description of scale based upon the expected difference between two randomly selected observations:

$$\lambda_2 = (1/2)\, E[X_{(2|1)} - X_{(1|2)}]$$

To compute L-moment measures of skewness three randomly selected observations are used and for kurtosis, we use four randomly selected observations.

$$\lambda_3 = (1/3)\, E[X_{(3|3)} - 2X_{(2|3)} + X_{(1|3)}]$$
$$\lambda_4 = (1/4)\, E[X_{(4|4)} - 3X_{(3|4)} + 3X_{(2|4)} - X_{(1|4)}]$$

Sample L-moment estimates are often computed using (PWMs). The rth PWM is defined (Loucks and Beek, 2005) as:

$$\beta_r = E\{X\,[F(X)]^r\}$$

where $F(X)$ is the cumulative distribution function of X. Recommended (Landwehr et al., 1979; Hosking and Wallis, 1995) unbiased PWM estimators, b_r, of β_r are computed as:

$$b_0 = \overline{X}$$

$$b_1 = \frac{1}{n(n-1)} \sum_{j=2}^{n} (j-1) X_{(j)}$$

$$b_2 = \frac{1}{n(n-1)(n-2)} \sum_{j=3}^{n} (j-1)(j-2) X_{(j)}$$

$$b_3 = \frac{1}{n(n-1)(n-2)(n-3)} \sum_{j=4}^{n} (j-1)(j-2)(j-3) X_{(j)}$$

The general formula for computing estimators b_r of β_r is given by

$$b_r = \frac{1}{n} \sum_{j=r+1}^{n} \binom{j-1}{r} X_{(j)} \Big/ \binom{n-1}{r}$$

$$= \frac{1}{r+1} \sum_{j=r+1}^{n} \binom{j-1}{r} X_{(j)} \Big/ \binom{n}{r+1}$$

for $r = 1, \ldots, n-1$.

L-moments are easily calculated in terms of PWMs using:

$$\lambda_1 = \beta_0$$
$$\lambda_2 = 2\beta_1 - \beta_0$$
$$\lambda_3 = 6\beta_2 - 6\beta_1 + \beta_0$$
$$\lambda_4 = 20\beta_3 - 30\beta_2 + 12\beta_1 - \beta_0$$

As with traditional product moments, measures of the coefficient of variation, skewness and kurtosis of a distribution can be computed with L-moments. Following L-moment ratios are important:

L- coefficient of variation (L-CV)	$t_2 = \lambda_2 / \lambda_1$
L- coefficient of skewness (L-sk)	$t_3 = \lambda_3 / \lambda_2$
L- coefficient of kurtosis (L-ku)	$t_4 = \lambda_4 / \lambda_2$

Example: Table gives annual discharge data of a river for 36 years. Compute sample L-moments and L-moment ratios, L-CV, L-sk, and L-ku.

Table: Annual discharge data of a river for 36 years

Year	Discharge	Year	Discharge	Year	Discharge	Year	Discharge
1950	400	1959	1390	1968	2291	1977	1499
1951	1100	1960	3300	1969	1340	1978	2598
1952	900	1961	2190	1970	3200	1979	3487
1953	440	1962	935	1971	2200	1980	1234
1954	3000	1963	785	1972	1014	1981	819
1955	2500	1964	501	1973	1790	1982	1210
1956	760	1965	1123	1974	1140	1983	1510
1957	1250	1966	1581	1975	764	1984	1780
1958	1340	1967	959	1976	783	1985	1398

Solution: Yields estimates of the first three Probability Weighted Moments:

$$b_0 = 1514.19$$
$$b_1 = 889.16$$
$$b_2 = 655.38$$
$$b_3 = 518.64$$

The sample L-moments can be calculated using the probability weighted moments to obtain:

$$\hat{\lambda}_1 = b_0 = 1514.19$$

$$\hat{\lambda}_2 = 2b_1 - b_0 = 264.12$$

$$\hat{\lambda}_3 = 6b_2 - 6b_1 + b_0 = 111.53$$

$$\hat{\lambda}_4 = 20b_3 - 30b_2 + 12b_1 - b_0 = -132.82$$

Thus, the sample estimates of the L-coefficient of variation, t_2, and L-coefficient of skewness, t_3, are:

$$t_2 = 264.12/1514.19 = 0.174$$
$$t_3 = 111.53/264.12 = 0.422$$
$$t_4 = -132.82/264.12 = -0.502$$

Problems of Parameter Estimation

The parameters of a distribution function are estimated from the available sample data. But while doing so, errors may arise due to many reasons. The sample data may contain errors, the assumption underlying a particular method of parameter estimation may not hold good, and there may be truncation and round-off errors. All these may result in errors in estimates of parameter. Each estimate of a parameter is a function of sample parameter data which are observations of a random variable. Thus, the estimate value of the parameter itself is a random variable with certain distribution. An estimate obtained from a given set of values can be regarded as an observed value of the random variable. Thus, the goodness of an estimate can be judged from its distribution.

Several questions arise in parameter estimation. How should we employ the available data to obtain the best estimate? What is the best estimate? Are these estimates unique? A number of statistical properties are available by which to address the above questions. These are discussed below.

Bias

Let the estimate of parameter a be a_c denoted by. Estimate a_c will be called an unbiased estimate of a if the expected value of a, denoted $E\,(a_c) = a$. In general, an estimate will have a certain bias $b(a)$ depending on a so that

$$E(a_c) = a + b$$

An unbiased estimate mean $b(a) = o$. Note that an individual a_c may not be equal to or close to a even if $b(a) = o$. Unbiasedness simply implies that the average of many independent estimates of a will be equal to a.

The bias in a given quantity is usually measured in dimensionless terms and is often referred to as standardized bias (or BIAS). Thus, BIAS is defined as

$$BIAS = \frac{E(\hat{a}) - a}{a}$$

where $\hat{a}$ is an estimate of parameter or quantile of a. In Monte Carlo experimentation, large numbers of samples of different sizes are generated from a given population. For each sample, then, an estimate of a is obtained. If there are, say, 1000 samples of a given size generated then there are 1000 values of parameter a. Thus, earlier equation is the average of the 1000 estimates of a for a given sample size and is estimated as

$$E(\hat{a}) = \sum_{i=1}^{n} \hat{a}_i / n$$

where n is the number of samples generated or the number of values of the a estimate. The value of a in eq. $BIAS = \frac{E(\hat{a}) - a}{a}$ is the true value of a or the value of parameter a of the population.

Efficiency

An estimate ac of a is said to be efficient if it is unbiased and its variance is at least as small as that of any other unbiased estimate of a. If there are two estimates of a, say a_1 and a_2, then the relative efficiency of a_1 with respect to a_2 is defined as

$$e = \frac{E(a_1 - a)^2}{E(a_2 - a)^2} \leq 1$$

if $E(a_2-a)^2 > E(a_1-a)^2$, then $e \leq 1$. An efficient estimate has e = 1. If an efficient estimate exists, it may be approximately obtained by use of the MLE or entropy method.

Standard Error

Another dimensionless performance measure frequently used in hydrology is the standard error (SE), defined as

$$SE = \sigma(\hat{a}) / a$$

where σ (.) denotes the standard deviation of a and is computed as

$$\sigma(\hat{a}) = \left[\frac{1}{n-1} \sum_{i=1}^{n} \{ \hat{a}_i - E(\hat{a}_i) \}^2 \right]^{1/2}$$

where the summations are over n estimates $\hat{a}$ of a. In Monte Carlo experiments, referred to as above, for each sample size, a value of SE is obtained. Thus, this measure is similar to the coefficient of variation.

Root Mean Square Error

The root mean square error (RMSE) is one of the most frequently employed performance measures and is defined for parameter a estimate as

$$RMSE = E\left[(\hat{a} - a)^2 \right]^{1/2} / a$$

where E[.] is the expectation of [.]. It can be shown that RMSE is related to BIAS and SE as

$$RMSE = \left[\frac{n-1}{n} SE^2 + BIAS^2 \right]^{1/2}$$

Relative Mean Error

Another measure of error in assessing the goodness of fit of hydrologic models is the relative mean error (RME) defined as

$$RME = \frac{1}{N}\left(\sum_{i=1}^{N}\left[\frac{Q_0 - Q_c}{Q_0}\right]^2\right)^{0.5}$$

in which N is the sample size, Q is the observed quantity of a given probability and Q_c is the computed quantity of the same probability. Also, used sometimes is the relative absolute error defined as

$$RAE = \frac{1}{N}\sum_{i=1}^{N}\left|\frac{Q_0 - Q_c}{Q_c}\right|$$

Statistical Hypothesis Testing

A statistical hypothesis, sometimes called confirmatory data analysis, is a hypothesis that is testable on the basis of observing a process that is modeled via a set of random variables. A statistical hypothesis test is a method of statistical inference. Commonly, two statistical data sets are compared, or a data set obtained by sampling is compared against a synthetic data set from an idealized model. A hypothesis is proposed for the statistical relationship between the two data sets, and this is compared as an alternative to an idealized null hypothesis that proposes no relationship between two data sets. The comparison is deemed *statistically significant* if the relationship between the data sets would be an unlikely realization of the null hypothesis according to a threshold probability—the significance level. Hypothesis tests are used in determining what outcomes of a study would lead to a rejection of the null hypothesis for a pre-specified level of significance. The process of distinguishing between the null hypothesis and the alternative hypothesis is aided by identifying two conceptual types of errors (type 1 & type 2), and by specifying parametric limits on e.g. how much type 1 error will be permitted.

An alternative framework for statistical hypothesis testing is to specify a set of statistical models, one for each candidate hypothesis, and then use model selection techniques to choose the most appropriate model. The most common selection techniques are based on either Akaike information criterion or Bayes factor.

Confirmatory data analysis can be contrasted with exploratory data analysis, which may not have pre-specified hypotheses.

Variations and Sub-classes

Statistical hypothesis testing is a key technique of both frequentist inference and Bayesian inference, although the two types of inference have notable differences. Statistical hypothesis tests define a procedure that controls (fixes) the probability of incorrectly *deciding* that a default position (null hypothesis) is incorrect. The procedure is based on how likely it would be for a set of observations to occur if the null hypothesis were true. Note that this probability of making an incorrect decision is *not* the probability that the null hypothesis is true, nor whether any specific alternative hypothesis is true. This contrasts with other possible techniques of decision theory in which the null and alternative hypothesis are treated on a more equal basis.

One naïve Bayesian approach to hypothesis testing is to base decisions on the posterior probability, but this fails when comparing point and continuous hypotheses. Other approaches to decision making, such as Bayesian decision theory, attempt to balance the consequences of incorrect decisions across all possibilities, rather than concentrating on a single null hypothesis. A number of other approaches to reaching a decision based on data are available via decision theory and optimal decisions, some of which have desirable properties. Hypothesis testing, though, is a dominant approach to data analysis in many fields of science. Extensions to the theory of hypothesis testing include the study of the power of tests, i.e. the probability of correctly rejecting the null hypothesis given that it is false. Such considerations can be used for the purpose of sample size determination prior to the collection of data.

The Testing Process

In the statistics literature, statistical hypothesis testing plays a fundamental role. The usual line of reasoning is as follows:

1. There is an initial research hypothesis of which the truth is unknown.

2. The first step is to state the relevant null and alternative hypotheses. This is important, as mis-stating the hypotheses will muddy the rest of the process.

3. The second step is to consider the statistical assumptions being made about the sample in doing the test; for example, assumptions about the statistical independence or about the form of the distributions of the observations. This is equally important as invalid assumptions will mean that the results of the test are invalid.

4. Decide which test is appropriate, and state the relevant test statistic T.

5. Derive the distribution of the test statistic under the null hypothesis from the assumptions. In standard cases this will be a well-known result. For example, the test statistic might follow a Student's t distribution or a normal distribution.

6. Select a significance level (α), a probability threshold below which the null hypothesis will be rejected. Common values are 5% and 1%.

7. The distribution of the test statistic under the null hypothesis partitions the possible values of T into those for which the null hypothesis is rejected—the so-called *critical region*—and those for which it is not. The probability of the critical region is α.

8. Compute from the observations the observed value t_{obs} of the test statistic T.

9. Decide to either reject the null hypothesis in favor of the alternative or not reject it. The decision rule is to reject the null hypothesis H_o if the observed value t_{obs} is in the critical region, and to accept or "fail to reject" the hypothesis otherwise.

An alternative process is commonly used:

1. Compute from the observations the observed value t_{obs} of the test statistic T.

2. Calculate the p-value. This is the probability, under the null hypothesis, of sampling a test statistic at least as extreme as that which was observed.

3. Reject the null hypothesis, in favor of the alternative hypothesis, if and only if the p-value is less than the significance level (the selected probability) threshold.

The two processes are equivalent. The former process was advantageous in the past when only tables of test statistics at common probability thresholds were available. It allowed a decision to be made without the calculation of a probability. It was adequate for classwork and for operational use, but it was deficient for reporting results.

The latter process relied on extensive tables or on computational support not always available. The explicit calculation of a probability is useful for reporting. The calculations are now trivially performed with appropriate software.

The difference in the two processes applied to the Radioactive suitcase example (below):

- "The Geiger-counter reading is 10. The limit is 9. Check the suitcase."

- "The Geiger-counter reading is high; 97% of safe suitcases have lower readings. The limit is 95%. Check the suitcase."

The former report is adequate, the latter gives a more detailed explanation of the data and the reason why the suitcase is being checked.

It is important to note the difference between accepting the null hypothesis and simply failing to reject it. The "fail to reject" terminology highlights the fact that the null hypothesis is assumed to be true from the start of the test; if there is a lack of evidence against it, it simply continues to be assumed true. The phrase "accept the null hypothesis" may suggest it has been proved simply because it has not been disproved, a logical fallacy known as the argument from ignorance. Unless a test with particularly high power is used, the idea of "accepting" the null hypothesis may be dangerous. Nonetheless the terminology is prevalent throughout statistics, where the meaning actually intended is well understood.

The processes described here are perfectly adequate for computation. They seriously neglect the design of experiments considerations.

It is particularly critical that appropriate sample sizes be estimated before conducting the experiment.

The phrase "test of significance" was coined by statistician Ronald Fisher.

Interpretation

If the p-value is less than the required significance level (equivalently, if the observed test statistic is in the critical region), then we say the null hypothesis is rejected at the given level of significance. Rejection of the null hypothesis is a conclusion. This is like a "guilty" verdict in a criminal trial: the evidence is sufficient to reject innocence, thus proving guilt. We might accept the alternative hypothesis (and the research hypothesis).

If the p-value is *not* less than the required significance level (equivalently, if the observed test statistic is outside the critical region), then the test has no result. The evidence is insufficient to support a conclusion. (This is like a jury that fails to reach a verdict.) The researcher typically gives extra consideration to those cases where the p-value is close to the significance level.

Some people find it helpful to think of the hypothesis testing framework as analogous to a mathematical proof by contradiction.

In the Lady tasting tea example, Fisher required the Lady to properly categorize all of the cups of tea to justify the conclusion that the result was unlikely to result from chance. He defined the critical region as that case alone. The region was defined by a probability (that the null hypothesis was correct) of less than 5%.

Whether rejection of the null hypothesis truly justifies acceptance of the research hypothesis depends on the structure of the hypotheses. Rejecting the hypothesis that a large paw print originated from a bear does not immediately prove the existence of Bigfoot. Hypothesis testing emphasizes the rejection, which is based on a probability, rather than the acceptance, which requires extra steps of logic.

"The probability of rejecting the null hypothesis is a function of five factors: whether the test is one- or two tailed, the level of significance, the standard deviation, the amount of deviation from the null hypothesis, and the number of observations." These factors are a source of criticism; factors under the control of the experimenter/analyst give the results an appearance of subjectivity.

Use and Importance

Statistics are helpful in analyzing most collections of data. This is equally true of hypothesis testing which can justify conclusions even when no scientific theory exists. In the Lady tasting tea example, it was "obvious" that no difference existed between (milk poured into tea) and (tea poured into milk). The data contradicted the "obvious".

Real world applications of hypothesis testing include:

- Testing whether more men than women suffer from nightmares

- Establishing authorship of documents

- Evaluating the effect of the full moon on behavior

- Determining the range at which a bat can detect an insect by echo

- Deciding whether hospital carpeting results in more infections

- Selecting the best means to stop smoking

- Checking whether bumper stickers reflect car owner behavior

- Testing the claims of handwriting analysts

Statistical hypothesis testing plays an important role in the whole of statistics and in statistical inference. For example, Lehmann (1992) in a review of the fundamental paper by Neyman and Pearson (1933) says: "Nevertheless, despite their shortcomings, the new paradigm formulated in the 1933 paper, and the many developments carried out within its framework continue to play a central role in both the theory and practice of statistics and can be expected to do so in the foreseeable future".

Significance testing has been the favored statistical tool in some experimental social sciences (over 90% of articles in the *Journal of Applied Psychology* during the early 1990s). Other fields have favored the estimation of parameters (e.g., effect size). Significance testing is used as a substitute for the traditional comparison of predicted value and experimental result at the core of the scientific method. When theory is only capable of predicting the sign of a relationship, a directional (one-sided) hypothesis test can be configured so that only a statistically significant result supports theory. This form of theory appraisal is the most heavily criticized application of hypothesis testing.

Cautions

"If the government required statistical procedures to carry warning labels like those on drugs, most inference methods would have long labels indeed." This caution applies to hypothesis tests and alternatives to them.

The successful hypothesis test is associated with a probability and a type-I error rate. The conclusion *might* be wrong.

The conclusion of the test is only as solid as the sample upon which it is based. The design of the experiment is critical. A number of unexpected effects have been observed including:

- The clever Hans effect. A horse appeared to be capable of doing simple arithmetic.

- The Hawthorne effect. Industrial workers were more productive in better illumination, and most productive in worse.

- The placebo effect. Pills with no medically active ingredients were remarkably effective.

A statistical analysis of misleading data produces misleading conclusions. The issue of data quality can be more subtle. In forecasting for example, there is no agreement on a measure of forecast accuracy. In the absence of a consensus measurement, no decision based on measurements will be without controversy.

The book *How to Lie with Statistics* is the most popular book on statistics ever published. It does not much consider hypothesis testing, but its cautions are applicable, including: Many claims are made on the basis of samples too small to convince. If a report does not mention sample size, be doubtful.

Hypothesis testing acts as a filter of statistical conclusions; only those results meeting a probability threshold are publishable. Economics also acts as a publication filter; only those results favorable to the author and funding source may be submitted for publication. The impact of filtering on publication is termed publication bias. A related problem is that of multiple testing (sometimes linked to data mining), in which a variety of tests for a variety of possible effects are applied to a single data set and only those yielding a significant result are reported. These are often dealt with by using multiplicity correction procedures that control the family wise error rate (FWER) or the false discovery rate (FDR).

Those making critical decisions based on the results of a hypothesis test are prudent to look at the details rather than the conclusion alone. In the physical sciences most results are fully accepted only when independently confirmed. The general advice concerning statistics is, "Figures never lie, but liars figure".

Examples

Lady Tasting Tea

In a famous example of hypothesis testing, known as the *Lady tasting tea*, Dr. Muriel Bristol, a female colleague of Fisher claimed to be able to tell whether the tea or the milk was added first to a cup. Fisher proposed to give her eight cups, four of each variety, in random order. One could then ask what the probability was for her getting the number she got correct, but just by chance. The null hypothesis was that the Lady had no such ability. The test statistic was a simple count of the number of successes in selecting the 4 cups. The critical region was the single case of 4 successes of 4 possible based on a conventional probability criterion (< 5%; 1 of 70 ≈ 1.4%). Fisher asserted that no alternative hypothesis was (ever) required. The lady correctly identified every cup, which would be considered a statistically significant result.

Courtroom Trial

A statistical test procedure is comparable to a criminal trial; a defendant is considered not guilty as long as his or her guilt is not proven. The prosecutor tries to prove the guilt of the defendant. Only when there is enough charging evidence the defendant is convicted.

In the start of the procedure, there are two hypotheses H_0: "the defendant is not guilty", and H_1: "the defendant is guilty". The first one, H_0, is called the *null hypothesis*, and is for the time being accepted. The second one, H_1, is called the *alternative hypothesis*. It is the alternative hypothesis that one hopes to support.

The hypothesis of innocence is only rejected when an error is very unlikely, because one doesn't want to convict an innocent defendant. Such an error is called *error of the first kind* (i.e., the conviction of an innocent person), and the occurrence of this error is controlled to be rare. As a consequence of this asymmetric behaviour, the *error of the second kind* (acquitting a person who committed the crime), is often rather large.

	H_0 is true Truly not guilty	H_1 is true Truly guilty
Accept null hypothesis Acquittal	Right decision	Wrong decision Type II Error
Reject null hypothesis Conviction	Wrong decision Type I Error	Right decision

A criminal trial can be regarded as either or both of two decision processes: guilty vs not guilty or evidence vs a threshold ("beyond a reasonable doubt"). In one view, the defendant is judged; in the other view the performance of the prosecution (which bears the burden of proof) is judged. A hypothesis test can be regarded as either a judgment of a hypothesis or as a judgment of evidence.

Philosopher's Beans

The following example was produced by a philosopher describing scientific methods generations before hypothesis testing was formalized and popularized.

Few beans of this handful are white. Most beans in this bag are white. Therefore: Probably, these beans were taken from another bag. This is an hypothetical inference.

The beans in the bag are the population. The handful are the sample. The null hypothesis is that the sample originated from the population. The criterion for rejecting the null-hypothesis is the "obvious" difference in appearance (an informal difference in the mean). The interesting result is that consideration of a real population and a real sample produced an imaginary bag. The philosopher was considering logic rather than probability. To be a real statistical hypothesis test, this example requires the formalities of a probability calculation and a comparison of that probability to a standard.

A simple generalization of the example considers a mixed bag of beans and a handful that contain either very few or very many white beans. The generalization considers both extremes. It requires more calculations and more comparisons to arrive at a formal answer, but the core philosophy is unchanged; If the composition of the handful is greatly different from that of the bag, then the sample probably originated from another bag. The original example is termed a one-sided or a one-tailed test while the generalization is termed a two-sided or two-tailed test.

The statement also relies on the inference that the sampling was random. If someone had been picking through the bag to find white beans, then it would explain why the handful had so many white beans, and also explain why the number of white beans in the bag was depleted (although the bag is probably intended to be assumed much larger than one's hand).

Clairvoyant Card Game

A person (the subject) is tested for clairvoyance. He is shown the reverse of a randomly chosen playing card 25 times and asked which of the four suits it belongs to. The number of hits, or correct answers, is called X.

As we try to find evidence of his clairvoyance, for the time being the null hypothesis is that the person is not clairvoyant. The alternative is, of course: the person is (more or less) clairvoyant.

If the null hypothesis is valid, the only thing the test person can do is guess. For every card, the probability (relative frequency) of any single suit appearing is $1/4$. If the alternative is valid, the test subject will predict the suit correctly with probability greater than $1/4$. We will call the probability of guessing correctly p. The hypotheses, then, are:

- null hypothesis : $H_0 : p = \frac{1}{4}$ (just guessing)

and

- alternative hypothesis : $H_1 : p > \frac{1}{4}$ (true clairvoyant).

When the test subject correctly predicts all 25 cards, we will consider him clairvoyant, and reject the null hypothesis. Thus also with 24 or 23 hits. With only 5 or 6 hits, on the other hand, there is no cause to consider him so. But what about 12 hits, or 17 hits? What is the critical number, c, of hits, at which point we consider the subject to be clairvoyant? How do we determine the critical value c? It is obvious that with the choice c=25 (i.e. we only accept clairvoyance when all cards are predicted correctly) we're more critical than with c=10. In the first case almost no test subjects will be recognized to be clairvoyant, in the second case, a certain number will pass the test. In practice, one decides how critical one will be. That is, one decides how often one accepts an error of the first kind – a false positive, or Type I error. With c = 25 the probability of such an error is:

$$P(\text{reject } H_0 \mid H_0 \text{ is valid}) = P(X = 25 \mid p = \frac{1}{4}) = \left(\frac{1}{4}\right)^{25} \approx 10^{-15},$$

and hence, very small. The probability of a false positive is the probability of randomly guessing correctly all 25 times.

Being less critical, with c=10, gives:

$$P(\text{reject } H_0 \mid H_0 \text{ is valid}) = P(X \geq 10 \mid p = \frac{1}{4}) = \sum_{k=10}^{25} P(X = k \mid p = \frac{1}{4}) \approx 0.07.$$

Thus, c = 10 yields a much greater probability of false positive.

Before the test is actually performed, the maximum acceptable probability of a Type I error (α) is determined. Typically, values in the range of 1% to 5% are selected. (If the maximum acceptable error rate is zero, an infinite number of correct guesses is required.) Depending on this Type 1 error rate, the critical value c is calculated. For example, if we select an error rate of 1%, c is calculated thus:

$$P(\text{reject } H_0 \mid H_0 \text{ is valid}) = P(X \geq c \mid p = \frac{1}{4}) \leq 0.01.$$

From all the numbers c, with this property, we choose the smallest, in order to minimize the probability of a Type II error, a false negative. For the above example, we select: $c = 13$.

Radioactive Suitcase

As an example, consider determining whether a suitcase contains some radioactive material. Placed under a Geiger counter, it produces 10 counts per minute. The null hypothesis is that no radioactive material is in the suitcase and that all measured counts are due to ambient radioactivity typical of the surrounding air and harmless objects. We can then calculate how likely it is that we would observe 10 counts per minute if the null hypothesis were true. If the null hypothesis predicts (say) on average 9 counts per minute, then according to the Poisson distribution typical for radioactive decay there is about 41% chance of recording 10 or more counts. Thus we can say that the suitcase is compatible with the null hypothesis (this does not guarantee that there is no radioactive material, just that we don't have enough evidence to suggest there is). On the other hand, if the null hypothesis predicts 3 counts per minute (for which the Poisson distribution predicts only 0.1% chance of recording 10 or more counts) then the suitcase is not compatible with the null hypothesis, and there are likely other factors responsible to produce the measurements.

The test does not directly assert the presence of radioactive material. A *successful* test asserts that the claim of no radioactive material present is unlikely given the reading (and therefore ...). The double negative (disproving the null hypothesis) of the method is confusing, but using a counter-example to disprove is standard mathematical practice. The attraction of the method is its practicality. We know (from experience) the expected range of counts with only ambient radioactivity present, so we can say that a measurement is *unusually* large. Statistics just formalizes the intuitive by using numbers instead of adjectives. We probably do not know the characteristics of the radioactive suitcases; We just assume that they produce larger readings.

To slightly formalize intuition: Radioactivity is suspected if the Geiger-count with the suitcase is among or exceeds the greatest (5% or 1%) of the Geiger-counts made with ambient radiation alone. This makes no assumptions about the distribution of counts. Many ambient radiation observations are required to obtain good probability estimates for rare events.

The test described here is more fully the null-hypothesis statistical significance test. The null hypothesis represents what we would believe by default, before seeing any evidence. Statistical significance is a possible finding of the test, declared when the observed sample is unlikely to have occurred by chance if the null hypothesis were true. The name of the test describes its formulation and its possible outcome. One characteristic of the test is its crisp decision: to reject or not reject the null hypothesis. A calculated value is compared to a threshold, which is determined from the tolerable risk of error.

Definition of Terms

The following definitions are mainly based on the exposition in the book by Lehmann and Romano:

Statistical hypothesis

>   A statement about the parameters describing a population (not a sample).

Statistic

> A value calculated from a sample, often to summarize the sample for comparison purposes.

Simple hypothesis

> Any hypothesis which specifies the population distribution completely.

Composite hypothesis

> Any hypothesis which does *not* specify the population distribution completely.

Null hypothesis (H_o)

> A simple hypothesis associated with a contradiction to a theory one would like to prove.

Alternative hypothesis (H_1)

> A hypothesis (often composite) associated with a theory one would like to prove.

Statistical test

> A procedure whose inputs are samples and whose result is a hypothesis.

Region of acceptance

> The set of values of the test statistic for which we fail to reject the null hypothesis.

Region of rejection / Critical region

> The set of values of the test statistic for which the null hypothesis is rejected.

Critical value

> The threshold value delimiting the regions of acceptance and rejection for the test statistic.

Power of a test ($1 - \beta$)

> The test's probability of correctly rejecting the null hypothesis. The complement of the false negative rate, β. Power is termed sensitivity in biostatistics. ("This is a sensitive test. Because the result is negative, we can confidently say that the patient does not have the condition.") See sensitivity and specificity and Type I and type II errors for exhaustive definitions.

Size

> For simple hypotheses, this is the test's probability of *incorrectly* rejecting the null hypothesis. The false positive rate. For composite hypotheses this

is the supremum of the probability of rejecting the null hypothesis over all cases covered by the null hypothesis. The complement of the false positive rate is termed specificity in biostatistics. ("This is a specific test. Because the result is positive, we can confidently say that the patient has the condition.").

Significance level of a test (α)

It is the upper bound imposed on the size of a test. Its value is chosen by the statistician prior to looking at the data or choosing any particular test to be used. It is the maximum exposure to erroneously rejecting H_0 he/she is ready to accept. Testing H_0 at significance level α means testing H_0 with a test whose size does not exceed α. In most cases, one uses tests whose size is equal to the significance level.

p-value

The probability, assuming the null hypothesis is true, of observing a result at least as extreme as the test statistic.

Statistical significance test

A predecessor to the statistical hypothesis test. An experimental result was said to be statistically significant if a sample was sufficiently inconsistent with the (null) hypothesis. This was variously considered common sense, a pragmatic heuristic for identifying meaningful experimental results, a convention establishing a threshold of statistical evidence or a method for drawing conclusions from data. The statistical hypothesis test added mathematical rigor and philosophical consistency to the concept by making the alternative hypothesis explicit. The term is loosely used to describe the modern version which is now part of statistical hypothesis testing.

Conservative test

A test is conservative if, when constructed for a given nominal significance level, the true probability of *incorrectly* rejecting the null hypothesis is never greater than the nominal level.

Exact test

A test in which the significance level or critical value can be computed exactly, i.e., without any approximation. In some contexts this term is restricted to tests applied to categorical data and to permutation tests, in which computations are carried out by complete enumeration of all possible outcomes and their probabilities.

A statistical hypothesis test compares a test statistic (z or t for examples) to a threshold. The test statistic is based on optimality. For a fixed level of Type I error rate, use of these statistics minimizes Type II error rates (equivalent to maximizing power). The following terms describe tests in terms of such optimality:

Most powerful test

> For a given *size* or *significance level*, the test with the greatest power (probability of rejection) for a given value of the parameter(s) being tested, contained in the alternative hypothesis.

Uniformly most powerful test (UMP)

> A test with the greatest *power* for all values of the parameter(s) being tested, contained in the alternative hypothesis.

Common Test Statistics

One-sample tests are appropriate when a sample is being compared to the population from a hypothesis. The population characteristics are known from theory or are calculated from the population.

Two-sample tests are appropriate for comparing two samples, typically experimental and control samples from a scientifically controlled experiment.

Paired tests are appropriate for comparing two samples where it is impossible to control important variables. Rather than comparing two sets, members are paired between samples so the difference between the members becomes the sample. Typically the mean of the differences is then compared to zero. The common example scenario for when a paired difference test is appropriate is when a single set of test subjects has something applied to them and the test is intended to check for an effect.

Z-tests are appropriate for comparing means under stringent conditions regarding normality and a known standard deviation.

A *t*-test is appropriate for comparing means under relaxed conditions (less is assumed).

Tests of proportions are analogous to tests of means (the 50% proportion).

Chi-squared tests use the same calculations and the same probability distribution for different applications:

- Chi-squared tests for variance are used to determine whether a normal population has a specified variance. The null hypothesis is that it does.

- Chi-squared tests of independence are used for deciding whether two variables are associated or are independent. The variables are categorical rather than numeric. It can be used to decide whether left-handedness is correlated with libertarian politics (or not). The null hypothesis is that the variables are independent. The numbers used in the calculation are the observed and expected frequencies of occurrence.

- Chi-squared goodness of fit tests are used to determine the adequacy of curves fit to data. The null hypothesis is that the curve fit is adequate. It is common to determine curve shapes to minimize the mean square error, so it is appropriate that the goodness-of-fit calculation sums the squared errors.

F-tests (analysis of variance, ANOVA) are commonly used when deciding whether groupings of data by category are meaningful. If the variance of test scores of the left-handed in a class is much smaller than the variance of the whole class, then it may be useful to study lefties as a group. The null hypothesis is that two variances are the same – so the proposed grouping is not meaningful.

In the table below, the symbols used are defined at the bottom of the table. Proofs exist that the test statistics are appropriate.

Name	Formula	Assumptions or notes
One-sample z-test	$$z = \frac{\bar{x} - \mu_0}{(\sigma / \sqrt{n})}$$	(Normal population or $n > 30$) and σ known. (z is the distance from the mean in relation to the standard deviation of the mean). For non-normal distributions it is possible to calculate a minimum proportion of a population that falls within k standard deviations for any k.
Two-sample z-test	$$z = \frac{(\bar{x}_1 - \bar{x}_2) - d_0}{\sqrt{\dfrac{\sigma_1^2}{n_1} + \dfrac{\sigma_2^2}{n_2}}}$$	Normal population and independent observations and σ_1 and σ_2 are known
One-sample t-test	$$t = \frac{\bar{x} - \mu_0}{(s / \sqrt{n})},$$ $$df = n - 1$$	(Normal population or $n > 30$) and σ unknown
Paired t-test	$$t = \frac{\bar{d} - d_0}{(s_d / \sqrt{n})},$$ $$df = n - 1$$	(Normal population of differences **or** $n > 30$) and σ unknown or small sample size $n < 30$

Two-sample pooled t-test, equal variances	$$t = \frac{(\bar{x}_1 - \bar{x}_2) - d_0}{s_p\sqrt{\frac{1}{n_1} + \frac{1}{n_2}}},$$ $$s_p^2 = \frac{(n_1 - 1)s_1^2 + (n_2 - 1)s_2^2}{n_1 + n_2 - 2},$$ $$df = n_1 + n_2 - 2$$	(Normal populations or $n_1 + n_2 > 40$) and independent observations and $\sigma_1 = \sigma_2$ unknown
Two-sample unpooled t-test, unequal variances (Welch's t-test)	$$t = \frac{(\bar{x}_1 - \bar{x}_2) - d_0}{\sqrt{\frac{s_1^2}{n_1} + \frac{s_2^2}{n_2}}},$$ $$df = \frac{\left(\frac{s_1^2}{n_1} + \frac{s_2^2}{n_2}\right)^2}{\frac{\left(\frac{s_1^2}{n_1}\right)^2}{n_1 - 1} + \frac{\left(\frac{s_2^2}{n_2}\right)^2}{n_2 - 1}}$$	(Normal populations or $n_1 + n_2 > 40$) and independent observations and $\sigma_1 \neq \sigma_2$ both unknown
One-proportion z-test	$$z = \frac{\hat{p} - p_0}{\sqrt{p_0(1 - p_0)}}\sqrt{n}$$	$n \cdot p_o > 10$ and $n(1 - p_o) > 10$ and it is a SRS (Simple Random Sample).
Two-proportion z-test, pooled for $H_0 : p_1 = p_2$	$$z = \frac{(\hat{p}_1 - \hat{p}_2)}{\sqrt{\hat{p}(1 - \hat{p})(\frac{1}{n_1} + \frac{1}{n_2})}}$$ $$\hat{p} = \frac{x_1 + x_2}{n_1 + n_2}$$	$n_1 p_1 > 5$ and $n_1(1 - p_1) > 5$ and $n_2 p_2 > 5$ and $n_2(1 - p_2) > 5$ and independent observations.
Two-proportion z-test, unpooled for $\lvert d_0 \rvert > 0$	$$z = \frac{(\hat{p}_1 - \hat{p}_2) - d_0}{\sqrt{\frac{\hat{p}_1(1 - \hat{p}_1)}{n_1} + \frac{\hat{p}_2(1 - \hat{p}_2)}{n_2}}}$$	$n_1 p_1 > 5$ and $n_1(1 - p_1) > 5$ and $n_2 p_2 > 5$ and $n_2(1 - p_2) > 5$ and independent observations.
Chi-squared test for variance	$$\chi^2 = (n - 1)\frac{s^2}{\sigma_0^2}$$	Normal population

Chi-squared test for goodness of fit	$\chi^2 = \sum\limits^{k} \dfrac{(\text{observed} - \text{expected})^2}{\text{expected}}$	$df = k - 1 - \#\ parameters\ estimated$, and one of these must hold. • All expected counts are at least 5. • All expected counts are > 1 and no more than 20% of expected counts are less than 5
Two-sample F test for equality of variances	$F = \dfrac{s_1^2}{s_2^2}$	Normal populations Arrange so $s_1^2 \geq s_2^2$ and reject H_o for $F > F(\alpha/2, n_1 - 1, n_2 - 1)$
Regression t-test of $H_0 : R^2 = 0.$	$t = \sqrt{\dfrac{R^2(n-k-1^*)}{1-R^2}}$	Reject H_o for $t > t(\alpha/2, n-k-1^*)$ *Subtract 1 for intercept; k terms contain independent variables.

In general, the subscript 0 indicates a value taken from the null hypothesis, H_o, which should be used as much as possible in constructing its test statistic. ... *Definitions of other symbols:*

- α, the probability of Type I error (rejecting a null hypothesis when it is in fact true)
- = sample size
- n_1 = sample 1 size
- n_2 = sample 2 size
- $\bar{x}$ = sample mean
- μ_0 = hypothesized population mean

- s^2 = sample variance
- s_1 = sample 1 standard deviation
- s_2 = sample 2 standard deviation
- t = t statistic
- df = degrees of freedom
- $\bar{d}$ = sample mean of differences

- $\hat{p}$ = x/n = sample proportion, unless specified otherwise
- P_0 = hypothesized population proportion
- P_1 = proportion 1
- P_2 = proportion 2
- d_p = hypothesized difference in proportion

- μ_1 = population 1 mean
- μ_2 = population 2 mean
- σ = population standard deviation
- σ^2 = population variance
- s = sample standard deviation
- $\sum\limits^{k}$ = sum (of k numbers)

- d_0 = hypothesized population mean difference
- s_d = standard deviation of differences
- χ^2 = Chi-squared statistic

- $\min\{n_1, n_2\}$ = minimum of n_1 and n_2
- $x_1 = n_1 p_1$
- $x_2 = n_2 p_2$
- F = F statistic

Origins and Early Controversy

Significance testing is largely the product of Karl Pearson (p-value, Pearson's chi-squared test), William Sealy Gosset (Student's t-distribution), and Ronald Fisher ("null hypothesis", analysis of variance, "significance test"), while hypothesis testing was developed by Jerzy Neyman and Egon Pearson (son of Karl). Ronald Fisher began his life in statistics as a Bayesian (Zabell 1992), but Fisher soon grew disenchanted with the subjectivity involved (namely use of the principle of indifference when determining prior probabilities), and sought to provide a more "objective" approach to inductive inference.

Fisher was an agricultural statistician who emphasized rigorous experimental design and methods to extract a result from few samples assuming Gaussian distributions. Neyman (who teamed with the younger Pearson) emphasized mathematical rigor and methods to obtain more results from many samples and a wider range of distributions. Modern hypothesis testing is an inconsistent hybrid of the Fisher vs Neyman/Pearson formulation, methods and terminology developed in the early 20th century. While hypothesis testing was popularized early in the 20th century, evidence of its use can be found much earlier. In the 1770s Laplace considered the statistics of almost half a million births. The statistics showed an excess of boys compared to girls. He concluded by calculation of a p-value that the excess was a real, but unexplained, effect.

Fisher popularized the "significance test". He required a null-hypothesis (corresponding to a population frequency distribution) and a sample. His (now familiar) calculations determined whether to reject the null-hypothesis or not. Significance testing did not utilize an alternative hypothesis so there was no concept of a Type II error.

The p-value was devised as an informal, but objective, index meant to help a researcher determine (based on other knowledge) whether to modify future experiments or strengthen one's faith in the null hypothesis. Hypothesis testing (and Type I/II errors) was devised by Neyman and Pearson as a more objective alternative to Fisher's p-value, also meant to determine researcher behaviour, but without requiring any inductive inference by the researcher.

Neyman & Pearson considered a different problem (which they called "hypothesis testing"). They initially considered two simple hypotheses (both with frequency distributions). They calculated two probabilities and typically selected the hypothesis associated with the higher probability (the hypothesis more likely to have generated the sample). Their method always selected a hypothesis. It also allowed the calculation of both types of error probabilities.

Fisher and Neyman/Pearson clashed bitterly. Neyman/Pearson considered their formulation to be an improved generalization of significance testing.(The defining paper was abstract. Mathematicians have generalized and refined the theory for decades.) Fisher thought that it was not applicable to scientific research because often, during the course of the experiment, it is discovered that the initial assumptions about the null hypothesis are questionable due to unexpected sources of error. He believed that the use of rigid reject/accept decisions based on models formulated before data is collected was incompatible with this common scenario faced by scientists and attempts to apply this method to scientific research would lead to mass confusion.

The dispute between Fisher and Neyman–Pearson was waged on philosophical grounds, characterized by a philosopher as a dispute over the proper role of models in statistical inference.

Events intervened: Neyman accepted a position in the western hemisphere, breaking his partnership with Pearson and separating disputants (who had occupied the same building) by much of the planetary diameter. World War II provided an intermission in the debate. The dispute between Fisher and Neyman terminated (unresolved after 27 years) with Fisher's death in 1962. Neyman wrote a well-regarded eulogy. Some of Neyman's later publications reported p-values and significance levels.

The modern version of hypothesis testing is a hybrid of the two approaches that resulted from confusion by writers of statistical textbooks (as predicted by Fisher) beginning in the 1940s. (But signal detection, for example, still uses the Neyman/Pearson formulation.) Great conceptual differences and many caveats in addition to those mentioned above were ignored. Neyman and Pearson provided the stronger terminology, the more rigorous mathematics and the more consistent philosophy, but the subject taught today in introductory statistics has more similarities with Fisher's method than theirs. This history explains the inconsistent terminology (example: the null hypothesis is never accepted, but there is a region of acceptance).

Sometime around 1940, in an apparent effort to provide researchers with a "non-controversial" way to have their cake and eat it too, the authors of statistical text books began anonymously combining these two strategies by using the p-value in place of the test statistic (or data) to test against the Neyman–Pearson "significance level". Thus, researchers were encouraged to infer the strength of their data against some null hypothesis using p-values, while also thinking they are retaining the post-data collection objectivity provided by hypothesis testing. It then became customary for the null hypothesis, which was originally some realistic research hypothesis, to be used almost solely as a strawman "nil" hypothesis (one where a treatment has no effect, regardless of the context).

A Comparison Between Fisherian, Frequentist (Neyman–Pearson)

Fisher's null hypothesis testing	Neyman–Pearson decision theory
1. Set up a statistical null hypothesis. The null need not be a nil hypothesis (i.e., zero difference).	1. Set up two statistical hypotheses, H1 and H2, and decide about α, β, and sample size before the experiment, based on subjective cost-benefit considerations. These define a rejection region for each hypothesis.
2. Report the exact level of significance (e.g., p = 0.051 or p = 0.049). Do not use a conventional 5% level, and do not talk about accepting or rejecting hypotheses. If the result is "not significant", draw no conclusions and make no decisions, but suspend judgement until further data is available.	2. If the data falls into the rejection region of H1, accept H2; otherwise accept H1. Note that accepting a hypothesis does not mean that you believe in it, but only that you act as if it were true.
3. Use this procedure only if little is known about the problem at hand, and only to draw provisional conclusions in the context of an attempt to understand the experimental situation.	3. The usefulness of the procedure is limited among others to situations where you have a disjunction of hypotheses (e.g., either $\mu_1 = 8$ or $\mu_2 = 10$ is true) and where you can make meaningful cost-benefit trade-offs for choosing alpha and beta.

Early Choices of Null Hypothesis

Paul Meehl has argued that the epistemological importance of the choice of null hypothesis has gone largely unacknowledged. When the null hypothesis is predicted by theory, a more precise experiment will be a more severe test of the underlying theory. When the null hypothesis defaults to "no difference" or "no effect", a more precise experiment is a less severe test of the theory that motivated performing the experiment. An examination of the origins of the latter practice may therefore be useful:

1778: Pierre Laplace compares the birthrates of boys and girls in multiple European cities. He states: "it is natural to conclude that these possibilities are very nearly in the same ratio". Thus Laplace's null hypothesis that the birthrates of boys and girls should be equal given "conventional wisdom".

1900: Karl Pearson develops the chi squared test to determine "whether a given form of frequency curve will effectively describe the samples drawn from a given population." Thus the null hypothesis is that a population is described by some distribution predicted by theory. He uses as an example the numbers of five and sixes in the Weldon dice throw data.

1904: Karl Pearson develops the concept of "contingency" in order to determine whether outcomes are independent of a given categorical factor. Here the null hypothesis is by default that two things are unrelated (e.g. scar formation and death rates from smallpox). The null hypothesis in this case is no longer predicted by theory or conventional wisdom, but is instead the principle of indifference that lead Fisher and others to dismiss the use of "inverse probabilities".

Null Hypothesis Statistical Significance Testing

An example of Neyman–Pearson hypothesis testing can be made by a change to the radioactive suitcase example. If the "suitcase" is actually a shielded container for the transportation of radioactive material, then a test might be used to select among three hypotheses: no radioactive source present, one present, two (all) present. The test could be required for safety, with actions required in each case. The Neyman–Pearson lemma of hypothesis testing says that a good criterion for the selection of hypotheses is the ratio of their probabilities (a likelihood ratio). A simple method of solution is to select the hypothesis with the highest probability for the Geiger counts observed. The typical result matches intuition: few counts imply no source, many counts imply two sources and intermediate counts imply one source.

Neyman–Pearson theory can accommodate both prior probabilities and the costs of actions resulting from decisions. The former allows each test to consider the results of earlier tests (unlike Fisher's significance tests). The latter allows the consideration of economic issues as well as probabilities. A likelihood ratio remains a good criterion for selecting among hypotheses.

The two forms of hypothesis testing are based on different problem formulations. The original test is analogous to a true/false question; the Neyman–Pearson test is more like multiple choice. In the view of Tukey the former produces a conclusion on the basis of only strong evidence while the latter produces a decision on the basis of available evidence. While the two tests seem quite different both mathematically and philosophically, later developments lead to the opposite claim. Consider many tiny radioactive sources. The hypotheses become 0,1,2,3... grains of radioactive sand. There is little distinction between none or some radiation (Fisher) and 0 grains of radioactive sand versus all of the alternatives (Neyman–Pearson). The major Neyman–Pearson paper of 1933 also considered composite hypotheses (ones whose distribution includes an unknown parameter). An example proved the optimality of the (Student's) t-test, "there can be no better test for the hypothesis under consideration" (p 321). Neyman–Pearson theory was proving the optimality of Fisherian methods from its inception.

Fisher's significance testing has proven a popular flexible statistical tool in application with little mathematical growth potential. Neyman–Pearson hypothesis testing is claimed as a pillar of mathematical statistics, creating a new paradigm for the field. It also stimulated new applications in statistical process control, detection theory, decision theory and game theory. Both formulations have been successful, but the successes have been of a different character.

The dispute over formulations is unresolved. Science primarily uses Fisher's (slightly modified) formulation as taught in introductory statistics. Statisticians study Neyman–Pearson theory in graduate school. Mathematicians are proud of uniting the formulations. Philosophers consider them separately. Learned opinions deem the formulations variously competitive (Fisher vs Neyman), incompatible or complementary. The dispute has become more complex since Bayesian inference has achieved respectability.

The terminology is inconsistent. Hypothesis testing can mean any mixture of two formulations that both changed with time. Any discussion of significance testing vs hypothesis testing is doubly vulnerable to confusion.

Fisher thought that hypothesis testing was a useful strategy for performing industrial quality control, however, he strongly disagreed that hypothesis testing could be useful for scientists. Hypothesis testing provides a means of finding test statistics used in significance testing. The concept of power is useful in explaining the consequences of adjusting the significance level and is heavily used in sample size determination. The two methods remain philosophically distinct. They usually (but *not always*) produce the same mathematical answer. The preferred answer is context dependent. While the existing merger of Fisher and Neyman–Pearson theories has been heavily criticized, modifying the merger to achieve Bayesian goals has been considered.

Criticism

Criticism of statistical hypothesis testing fills volumes citing 300–400 primary references. Much of the criticism can be summarized by the following issues:

- The interpretation of a p-value is dependent upon stopping rule and definition of multiple comparison. The former often changes during the course of a study and the latter is unavoidably ambiguous. (i.e. "p values depend on both the (data) observed and on the other possible (data) that might have been observed but weren't").

- Confusion resulting (in part) from combining the methods of Fisher and Neyman–Pearson which are conceptually distinct.

- Emphasis on statistical significance to the exclusion of estimation and confirmation by repeated experiments.

- Rigidly requiring statistical significance as a criterion for publication, resulting in publication bias. Most of the criticism is indirect. Rather than being wrong, statistical hypothesis testing is misunderstood, overused and misused.

- When used to detect whether a difference exists between groups, a paradox arises. As improvements are made to experimental design (e.g., increased precision of measurement and sample size), the test becomes more lenient. Unless one accepts the absurd assumption that all sources of noise in the data cancel out completely, the chance of finding statistical significance in either direction approaches 100%.

- Layers of philosophical concerns. The probability of statistical significance is a function of decisions made by experimenters/analysts. If the decisions are based on convention they are termed arbitrary or mindless while those not so based may be termed subjective. To minimize type II errors, large samples are recommended. In psychology practically all null hypotheses are claimed to be false for sufficiently large samples so "...it is usually nonsensical to perform an experiment with the *sole* aim of rejecting the null hypothesis.". "Statistically significant findings are often misleading" in psychology. Statistical significance does not imply practical significance and correlation does not imply causation. Casting doubt on the null hypothesis is thus far from directly supporting the research hypothesis.

- It does not tell us what we want to know.

Critics and supporters are largely in factual agreement regarding the characteristics of null hypothesis significance testing (NHST): While it can provide critical information, it is *inadequate as the sole tool for statistical analysis. Successfully rejecting the null hypothesis may offer no support for the research hypothesis.* The continuing controversy concerns the selection of the best statistical practices for the near-term future given the (often poor) existing practices. Critics would prefer to ban NHST completely, forcing a complete departure from those practices, while supporters suggest a less absolute change.

Controversy over significance testing, and its effects on publication bias in particular, has produced several results. The American Psychological Association has strengthened its statistical reporting requirements after review, medical journal publishers have recognized the obligation to publish some results that are not statistically significant to combat publication bias and a journal (*Journal of Articles in Support of the Null Hypothesis*) has been created to publish such results exclusively. Textbooks have added some cautions and increased coverage of the tools necessary to estimate the size of the sample required to produce significant results. Major organizations have not abandoned use of significance tests although some have discussed doing so.

Alternatives

The numerous criticisms of significance testing do not lead to a single alternative. A unifying position of critics is that statistics should not lead to a conclusion or a decision but to a probability or to an estimated value with a confidence interval rather than to an accept-reject decision regarding a particular hypothesis. It is unlikely that the controversy surrounding significance testing will be resolved in the near future. Its supposed flaws and unpopularity do not eliminate the need for an objective and transparent means of reaching conclusions regarding studies that produce statistical results. Critics have not unified around an alternative. Other forms of reporting confidence or uncertainty could probably grow in popularity. One strong critic of significance testing suggested a list of reporting alternatives: effect sizes for importance, prediction intervals for confidence, replications and extensions for replicability, meta-analyses for generality. None of these suggested alternatives produces a conclusion/decision. Lehmann said that hypothesis testing theory can be presented in terms of conclusions/decisions, probabilities, or confidence intervals. "The distinction between the ... approaches is largely one of reporting and interpretation."

On one "alternative" there is no disagreement: Fisher himself said, "In relation to the test of significance, we may say that a phenomenon is experimentally demonstrable when we know how to conduct an experiment which will rarely fail to give us a statistically significant result." Cohen, an influential critic of significance testing, concurred, "... don't look for a magic alternative to NHST *[null hypothesis significance testing]* ... It doesn't exist." "... given the problems of statistical induction, we must finally rely, as have the older sciences, on replication." The "alternative" to significance testing is repeated testing. The easiest way to decrease statistical uncertainty is by obtaining more data, whether by increased sample size or by repeated tests. Nickerson claimed to have never seen the publication of a literally replicated experiment in psychology. An indirect approach to replication is meta-analysis.

Bayesian inference is one proposed alternative to significance testing. (Nickerson cited 10 sources suggesting it, including Rozeboom (1960)). For example, Bayesian parameter estimation can provide rich information about the data from which researchers can draw inferences, while using uncertain priors that exert only minimal influence on the results when enough data is available. Psychologist John K. Kruschke has suggested Bayesian estimation as an alternative for the *t*-test. Alternatively two competing models/hypothesis can be compared using Bayes factors. Bayesian methods could be criticized for requiring information that is seldom available in the cases where significance testing is most heavily used. Neither the prior probabilities nor the probability distribution of the test statistic under the alternative hypothesis are often available in the social sciences.

Advocates of a Bayesian approach sometimes claim that the goal of a researcher is most often to objectively assess the probability that a hypothesis is true based on the data they have collected. Neither Fisher's significance testing, nor Neyman–Pearson hypothesis testing can provide this information, and do not claim to. The probability a hypothesis is true can only be derived from use of Bayes' Theorem, which was unsatisfactory to both the Fisher and Neyman–Pearson camps due to the explicit use of subjectivity in the form of the prior probability. Fisher's strategy is to sidestep this with the p-value (an objective *index* based on the data alone) followed by *inductive inference*, while Neyman–Pearson devised their approach of *inductive behaviour*.

Philosophy

Hypothesis testing and philosophy intersect. Inferential statistics, which includes hypothesis testing, is applied probability. Both probability and its application are intertwined with philosophy. Philosopher David Hume wrote, "All knowledge degenerates into probability." Competing practical definitions of probability reflect philosophical differences. The most common application of hypothesis testing is in the scientific interpretation of experimental data, which is naturally studied by the philosophy of science.

Fisher and Neyman opposed the subjectivity of probability. Their views contributed to the objective definitions. The core of their historical disagreement was philosophical.

Many of the philosophical criticisms of hypothesis testing are discussed by statisticians in other contexts, particularly correlation does not imply causation and the design of experiments. Hypothesis testing is of continuing interest to philosophers.

Education

Statistics is increasingly being taught in schools with hypothesis testing being one of

the elements taught. Many conclusions reported in the popular press (political opinion polls to medical studies) are based on statistics. An informed public should understand the limitations of statistical conclusions and many college fields of study require
a course in statistics for the same reason. An introductory college statistics class places
much emphasis on hypothesis testing – perhaps half of the course. Such fields as literature and divinity now include findings based on statistical analysis. An introductory
statistics class teaches hypothesis testing as a cookbook process. Hypothesis testing is
also taught at the postgraduate level. Statisticians learn how to create good statistical
test procedures (like z, Student's t, F and chi-squared). Statistical hypothesis testing is
considered a mature area within statistics, but a limited amount of development continues.

The cookbook method of teaching introductory statistics leaves no time for history, philosophy or controversy. Hypothesis testing has been taught as received unified method.
Surveys showed that graduates of the class were filled with philosophical misconceptions (on all aspects of statistical inference) that persisted among instructors. While the
problem was addressed more than a decade ago, and calls for educational reform continue, students still graduate from statistics classes holding fundamental misconceptions
about hypothesis testing. Ideas for improving the teaching of hypothesis testing include
encouraging students to search for statistical errors in published papers, teaching the
history of statistics and emphasizing the controversy in a generally dry subject.

References

- Harlow, Lisa Lavoie; Stanley A. Mulaik; James H. Steiger, eds. (1997). What If There Were No Significance Tests?. Lawrence Erlbaum Associates. ISBN 978-0-8058-2634-0

- Fisher, R (1955). "Statistical Methods and Scientific Induction" (PDF). Journal of the Royal Statistical Society, Series B. 17 (1): 69–78

- B. S. Everitt: The Cambridge Dictionary of Statistics, Cambridge University Press, Cambridge (3rd edition, 2006). ISBN 0-521-69027-7

- C. S. Peirce (August 1878). "Illustrations of the Logic of Science VI: Deduction, Induction, and Hypothesis". Popular Science Monthly. 13. Retrieved March 30, 2012

- Gigerenzer, G (November 2004). "Mindless statistics". The Journal of Socio-Economics. 33 (5): 587–606. doi:10.1016/j.socec.2004.09.033

- Burnham, K. P.; Anderson, D. R. (2002). Model Selection and Multimodel Inference: A Practical Information-Theoretic Approach (2nd ed.). Springer-Verlag. ISBN 0-387-95364-7

- Cornfield, Jerome (1976). "Recent Methodological Contributions to Clinical Trials" (PDF). American Journal of Epidemiology. 104 (4): 408–421

- Lehmann, E. L.; Romano, Joseph P. (2005). Testing Statistical Hypotheses (3E ed.). New York: Springer. ISBN 0-387-98864-5

- Loveland, Jennifer L. (2011). Mathematical Justification of Introductory Hypothesis Tests and Development of Reference Materials (M.Sc. (Mathematics)). Utah State University. Retrieved April 30, 2013

- Begg, Colin B.; Berlin, Jesse A. (1988). "Publication bias: a problem in interpreting medical data". Journal of the Royal Statistical Society, Series A: 419–463

- Hinkelmann, Klaus and Kempthorne, Oscar (2008). Design and Analysis of Experiments. I and II (Second ed.). Wiley. ISBN 978-0-470-38551-7

- Kruschke, J K (July 9, 2012). "Bayesian Estimation Supersedes the T Test". Journal of Experimental Psychology: General. 142: 573–603. doi:10.1037/a0029146

- Richard J. Larsen; Donna Fox Stroup (1976). Statistics in the Real World: a book of examples. Macmillan. ISBN 978-0023677205

- Armstrong, J. Scott (2007). "Significance tests harm progress in forecasting". International Journal of Forecasting. 23 (2): 321–327. doi:10.1016/j.ijforecast.2007.03.004

Correlation and Regression Analysis

The connection between hydrologic variables can best be understood through cause and effect. If a change occurs in one variable, the other variable also changes. Regression analysis is the statistical method of measuring the relationship between variables. The topics discussed in the section are of great importance to broaden the existing knowledge on the subject matter.

Correlation Analysis

Many hydrologic variables are related to each other through cause and effect – changes in the values of one or more variables cause changes in some other variable. When simultaneous observations on such hydrological variables are available, one may be interested in finding out how strong is such association. Linear association between hydrologic variables is expressed by the covariance and correlation.

Let there be N pairs of observations (x_1, y_1), (x_2, y_2), ..., (x_N, y_N), of two hydrologic variables X and Y. The (sample) covariance between the variables is obtained from the following:

$$C_{XY} = \frac{1}{N-1} \sum_{i=1}^{N} (X_i - m_X)(y_i - m_Y)$$

where m_X and m_Y are mean values of X and Y respectively. Covariance is basically a measure of how much two variables under consideration change together. It is sometimes called a measure of "linear dependence" between two variables. In a special case when the two variables are identical, the covariance becomes variance.

Correlation is a measure of the strength of relationship between two variables or within the same series. Normally a dimensionless coefficient called correlation coefficient is computed which is a mathematical measure of the strenght of linear relationship. Note that correlation coefficient alone is not an evidence of a causal relationship between two variables. A set of variables may be releted due to two reasons. If one variable drives the other, they may be correlated, as rainfall and runoff. The variables may also be correlated if they share the same cause. Examples include dependent variables, such as river discharge, concentration or transport rates of sediment, and concentration or transport rates of substances that are transported in association with suspended sediment.

The correlation coefficient r_{XY} is obtained by scaling the covariance by the standard deviations of X and Y:

$$XY = \frac{XY}{s_X s_Y} = \frac{\overline{\sum(x_i - m_X)(y_i - m_Y)}}{s_X s_Y}$$

where s_x, s_Y = standard deviations of X and Y.

Since covariance depends on the units of the data, it is not possible to compare covariances among data sets having variables of different units or scales. This difficulty is addressed in correlation coefficient which is dimensionless quantity created by normalizing the covariance with the help of the product of the standard deviations of the variables.

Correlation oefficient measures linear association between the variables; non-linear association is not accounted for by this coefficient. Hence, if there is no linear association between X and Y then r_{XY} will be 0. Figure shows the various cases of association between two variables. The top left panel shows a case in which the two variables have a strong linear correlation and in this case r is likely to be close to 1. The top right panel shows a case when there is no correlation between the two variable and they are independent. In this case r is likely to be close to 0. An interesting situation is shown in the bottom right panel where the two variables do not have linear association and the correlation coefficient will be close to zero. But this does not necessarily mean that the variables X and Y are independent or there is no association between them. It only means that there is no linear association and there may be a circular association as shown here. Scatter plot of the data is a powerful and simple means to find out the presence of association between two variables.

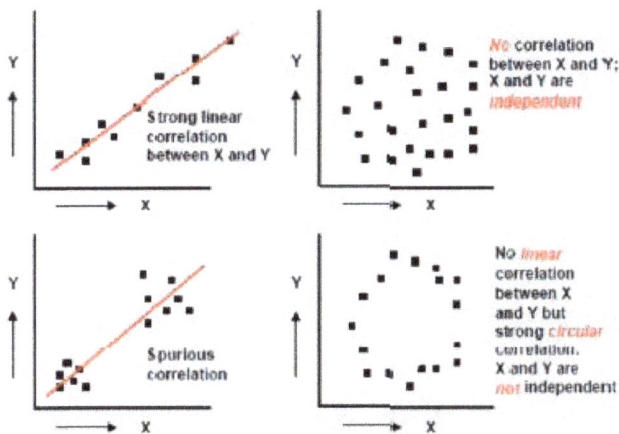

Various cases of association between two variables.

Since correlation coefficient is scaled, it can vary only within a certain range. To determine this range, assume that X and Y have a perfect linear association. Hence the relationship between X and Y can be given by:

Y = a + bX and on this basis one can write:

$$m_y = a + bm_x \quad and \quad s_y^2 = b^2 s_x^2 \quad or \quad s_y = |b| s_x$$

Substituting above relations in gives:

$$r_{XY} = \frac{\frac{1}{N-1}\sum (x_i - m_X)\left[a + bx_i - (a + bm_x)\right]}{S_X |b| S_x} = \frac{b}{|b|} \frac{\frac{1}{N-1}\sum (x_i - m_x)^2}{S_x^2} = \frac{b}{|b|}$$

Thus one can see that r_{XY} can vary between -1 to +1 or $-1 \le r_{XY} \le 1$. If X and Y are positively correlated then b > 0 and with increasing X, Y will also increase. In this case, r_{XY} will be near 1. If, on the other hand, X and Y are negatively correlated then b < 0 and Y will decrease when X increases. In such cases, r_{XY} will be near to −1.

If the scatter plot indicates a non-linear relationship between the two variables, one can try to transform the data by using, for example, logarithmic, square root, etc. transformations. In many cases, transformed data show better linear correlation and helps in drawing useful inferences.

Serial or Auto-Correlation

Consider that we have a time-series of data whose observations are equally spaced in time and that the statistical properties of the process do not change with time. The autocorrelation (correlation with self) or serial correlation of a series is the linear correlation between a time series and the values of the same series at a later interval of time. For example, this could be the correlation between the series $(x_1, x_2, ... x_n)$ and $(x_{1+k}, x_{2+k}, ... x_{n+k})$. Here 'k' is an integer termed as lag which may have values 1, 2, ... The autocorrelation of a time series (having n observations) at lag k (r_k) is given by

$$r_k = \frac{[\sum_{i=1}^{n-k} x_i x_{i+k} - \sum_{i=1}^{n-k} x_i \sum_{i=1}^{n-k} x_{i+k}]/(n-k)}{[\sum_{i=1}^{n-k} x_i^2 - (\sum_{i=1}^{n-k} x_i)^2/(n-k)]^{0.5}[\sum_{i=1}^{n-k} x_{i+k}^2 - (\sum_{i=1}^{n-k} x_{i+k})^2/(n-k)]^{0.5}}$$

Here the lag is the amount of offset when computing the autocorrelation. For example, the autocorrelation at lag 1 is determined by computing the correlation between elements 1, 2, ...,(n-1) of a series and the elements 2, 3, 4, ... n of the same series. From equation, it is clear that r_o is unity. Note that as k increases, the number of pairs of observations used in estimating r_k decreases since the summations contain $(n - k)$ terms. Therefore, serial correlation should only be estimated for k sufficiently smaller than n; usually correlation at lags exceeding 20 are not much useful.

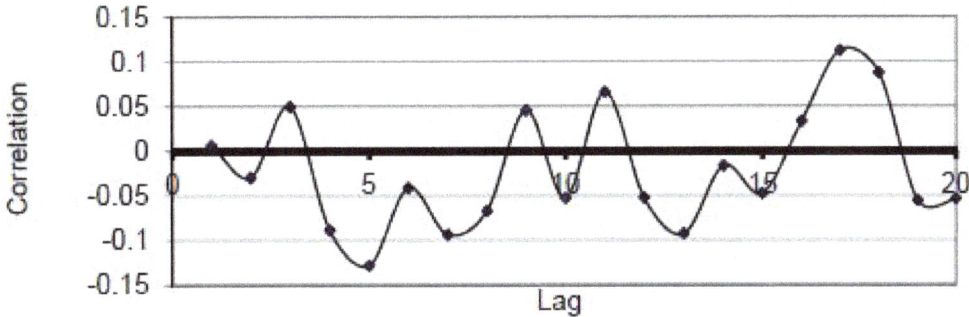

Autocorrelation r_k will be zero for all k for a pure random time series which implies that all the observations in the series are independent of each other. Conversely, if a time series of data shows high serial correlation then its elements cannot be termed as random elements.

A plot of autocorrelations at different lags is known as a correlogram. Typically correlogram begins at a value +1.0 at lag zero and the values decay at higher lags. Correlograms help reveal the characteristics of a time-series and disclose intervals of time or lags at which the time series has a repetitive nature. For a cyclic series, for example, monthly river flows, correlogram will also be cyclic.

The correlogram of the annual flows of Sabarmati River at Dharoi have been plotted up to 20 lags in figure. It can be seen from the correlogram that there is very poor autocorrelation in the series. The auto-correlation at lag 1 is 0.006 (nearly zero) and at lag 2, it is $- 0.0295$.

Cross-Correlation

Correlation between two variables x and y is also known as cross-correlation. As with autocorrelation, $r_{x,y}$ can also vary in the range from -1 to 1; r_{xy} = +1 or -1 implies a perfect linear relationship between X and Y; $r_{x,y}$ = 0 implies linear independence, although there may be other types (say, non-linear) of dependence. If observations in a time-series are correlated, this must be kept in mind while drawing any inferences about the data or when modeling the process that has produced the time-series.

A high correlation between two variables need not necessarily be due to a cause-and-effect relation between them. The monthly flows of two adjacent streams may be highly correlated but this could be because the influencing external causes are the same. A high correlation in this case does not mean that a change in flow of one stream will force the other stream's flow to change. It is to be noted that independent variables are uncorrelated but uncorrelated variables are not necessarily independent. The dependence in correlated variables is a stochastic dependence and not always physical or cause-and-effect dependence.

Spurious Correlation

Any apparent correlation between variables that are in fact uncorrelated or do not have a causeand- effect relation is termed as spurious correlation. One can arrive at wrong conclusions in hydrologic analysis on account spurious correlation if the physical reasons are overlooked. For example, the lower left panel in figure gives an example of spurious correlation. This type of situation may arise when data of two river gauging sites in different regions, both having distinct wet and dry periods are plotted. The data at X and Y may be uncorrelated in the wet period but the existence of dry and wet periods clusters observations in low and high flow ranges and the correlation appears to be very high.

One may also arrive at wrong conclusions by comparing data having the same denominator. For example, consider that X, Y and Z are uncorrelated and we attempt to try to find correlation between X/Z and Y/Z. Yevjevich (1972) showed that there will be a non-zero correlation, given by:

$$r = \frac{C_{V,Z}^2}{\left(C_{V,X}^2 + C_{V,Z}^2\right)^{1/2}\left(C_{V,Y}^2 + C_{V,Z}^2\right)^{1/2}}$$

It can be seen that when all the coefficients of variation are equal, r = 0.5! This example indicates that one needs to select the sample sets to be subjected to correlation and regression analysis carefully. Common divisors should be avoided.

Inferences on Correlation Coefficients

For uncorrleated variables, $r_{x,y}$ = 0 and for correlated ones, $r_{x,y} \neq 0$. However, even though there is no cause-and-effect and the data are not correlated, the correlation coefficient is rarely zero. This non-zero value of the correlation coefficent is likely to be due to chance. Thus, statistical tests are needed to ascertain if this deviation of the sample correlation coefficient from zero could be ascribed to chance or not. To conduct this statistical test, we need to make a null hypothesis that $r_{x,y}$ is zero or it has some known value. In other words, $H_o: r_{x,y}$ = 0 or $H_o: r_{x,y} = r^*$ where r^* is known. At the same time, an appropriate alternate hypothesis is also to be formulated.

Assume that X and Y are random variables from a bivariate normal distribution. If their correlation is 0, then the quantity

$$t = r[(n-2)/(1-r^2)]^{0.5}$$

follows a t-distribution with (n−2) degrees of freedom. The null hypothesis $H_o: r_{x,y} = 0$ is rejected if $|t| > t_{1-\alpha/2, n-2}$.

Kendall's Rank Correlation Test

It is also known as Kendall's τ test and is an effective and general test of correlation between two variables. Since it is a rank-based procedure, it overcomes the problems due to the effect of extreme values and the deviations from a linear relationship. Thus, this test is well-suited for use with dependent variables for which the variation around the general relationship exhibits a high degree of skewness or kurtosis. Examples include dependent variables such as river discharge, and concentration or transport rates of sediment.

The steps to conduct Kendall's test for correlation (the null hypothesis H_o is that the distribution of dependent variable y does not change as a function of independent variable x) are as follows:

1. The n pairs of data $(x_1, y_1), (x_2, y_2), \ldots (x_n, y_n)$ are indexed according to the magnitude of the x value, such that $x_1 \leq x_2 \leq \ldots \leq x_n$.

2. Compute the statistic S

$$S = \sum_{k=1}^{n-1} \sum_{j=k+1}^{n} sign\left(y_j - y_k\right)$$

    where

    sign $(\theta) = 1$ if $\theta > 0$

             $= 0$ if $\theta = 0$

             $= -1$ if $\theta < 0$.

3. For n > 10, the test is conducted using a normal approximation (Hirsch et al., 1993). The standardized test statistics Z is computed as:

$$Z = \begin{cases} \dfrac{S-1}{\sqrt{Var(S)}} & S > 0 \\ 0 & S = 0 \\ \dfrac{S+1}{\sqrt{Var(S)}} & S < 0 \end{cases}$$

where Var (S) = n(n − 1) (2n + 5)/18.

4. The null hypothesis is rejected at a significance level α if $|Z| > Z_{(1-\alpha/2)}$, where

$Z_{(1-\alpha/2)}$ is the value of the standard normal variate with a probability of exceedance of $\alpha/2$. If some of the x and/or y values are tied, the formula for Var (S) is modified. If the sample size is less than 10, then it is necessary to use tables for the S statistic.

Example: The correlation between the precipitation and runoff data (16 pairs of data were used) is 0.812. Test the null hypothesis H_0: the distribution of runoff does not change as a function of precipitation.

Solution: For this data, 16 pairs of values are available. These are arranged in ascending order. Using equation, the S statistic is found to be 72.

Hence, Var(S) = 16*(16-1)*(2*16+5)/18 = 493.33.

Now, applying eq.

$$Z = \left(72 - 1\right) / \sqrt{493.33} = 3.197$$

Since $|z| > 1.96$, the null hypothesis is rejected at 5% significance level and we can conclude that the distribution of runoff changes as a function of precipitation.

According to Yevjevich (1972), the 95% confidence limits for zero correlation are:

$$CL_{\pm}\left(k\right) = \frac{1}{n-k+1} \pm 1.96 \frac{\left(n-k-1\right)}{\left(n-k+1\right)} \sqrt{\frac{1}{n-k}}$$

Note that the confidence limits are not symmetrical about zero. The confidence region expands slightly with increasing lag. If the serial correlation for a particular lag falls within the confidence limits, then for that lag the serial correlation is not significant.

Regression Analysis

In statistical modeling, regression analysis is a statistical process for estimating the relationships among variables. It includes many techniques for modeling and analyzing several variables, when the focus is on the relationship between a dependent variable and one or more independent variables (or 'predictors'). More specifically, regression analysis helps one understand how the typical value of the dependent variable (or 'criterion variable') changes when any one of the independent variables is varied, while the other independent variables are held fixed. Most commonly, regression analysis estimates the conditional expectation of the dependent variable given the independent variables – that is, the average value of the dependent variable when the independent

variables are fixed. Less commonly, the focus is on a quantile, or other location parameter of the conditional distribution of the dependent variable given the independent
variables. In all cases, the estimation target is a function of the independent variables
called the regression function. In regression analysis, it is also of interest to characterize the variation of the dependent variable around the regression function which can
be described by a probability distribution. A related but distinct approach is necessary
condition analysis (NCA), which estimates the maximum (rather than average) value of
the dependent variable for a given value of the independent variable (ceiling line rather
than central line) in order to identify what value of the independent variable is necessary but not sufficient for a given value of the dependent variable.

Regression analysis is widely used for prediction and forecasting, where its use has
substantial overlap with the field of machine learning. Regression analysis is also used
to understand which among the independent variables are related to the dependent
variable, and to explore the forms of these relationships. In restricted circumstances,
regression analysis can be used to infer causal relationships between the independent
and dependent variables. However this can lead to illusions or false relationships, so
caution is advisable; for example, correlation does not imply causation.

Many techniques for carrying out regression analysis have been developed. Familiar
methods such as linear regression and ordinary least squares regression are parametric, in that the regression function is defined in terms of a finite number of unknown
parameters that are estimated from the data. Nonparametric regression refers to techniques that allow the regression function to lie in a specified set of functions, which may
be infinite-dimensional.

The performance of regression analysis methods in practice depends on the form of the
data generating process, and how it relates to the regression approach being used. Since
the true form of the data-generating process is generally not known, regression analysis often depends to some extent on making assumptions about this process. These
assumptions are sometimes testable if a sufficient quantity of data is available. Regression models for prediction are often useful even when the assumptions are moderately
violated, although they may not perform optimally. However, in many applications,
especially with small effects or questions of causality based on observational data, regression methods can give misleading results.

In a narrower sense, regression may refer specifically to the estimation of continuous
response variables, as opposed to the discrete response variables used in classification.
The case of a continuous output variable may be more specifically referred to as metric
regression to distinguish it from related problems.

History

The earliest form of regression was the method of least squares, which was published by
Legendre in 1805, and by Gauss in 1809. Legendre and Gauss both applied the method

to the problem of determining, from astronomical observations, the orbits of bodies about the Sun (mostly comets, but also later the then newly discovered minor planets). Gauss published a further development of the theory of least squares in 1821, including a version of the Gauss–Markov theorem.

The term "regression" was coined by Francis Galton in the nineteenth century to describe a biological phenomenon. The phenomenon was that the heights of descendants of tall ancestors tend to regress down towards a normal average (a phenomenon also known as regression toward the mean). For Galton, regression had only this biological meaning, but his work was later extended by Udny Yule and Karl Pearson to a more general statistical context. In the work of Yule and Pearson, the joint distribution of the response and explanatory variables is assumed to be Gaussian. This assumption was weakened by R.A. Fisher in his works of 1922 and 1925. Fisher assumed that the conditional distribution of the response variable is Gaussian, but the joint distribution need not be. In this respect, Fisher's assumption is closer to Gauss's formulation of 1821.

In the 1950s and 1960s, economists used electromechanical desk calculators to calculate regressions. Before 1970, it sometimes took up to 24 hours to receive the result from one regression.

Regression methods continue to be an area of active research. In recent decades, new methods have been developed for robust regression, regression involving correlated responses such as time series and growth curves, regression in which the predictor (independent variable) or response variables are curves, images, graphs, or other complex data objects, regression methods accommodating various types of missing data, nonparametric regression, Bayesian methods for regression, regression in which the predictor variables are measured with error, regression with more predictor variables than observations, and causal inference with regression.

Regression Models

Regression models involve the following variables:

- The unknown parameters, denoted as β, which may represent a scalar or a vector.

- The independent variables, X.

- The dependent variable, Y.

In various fields of application, different terminologies are used in place of dependent and independent variables.

A regression model relates Y to a function of X and β.

$$Y \approx f(X, \beta)$$

The approximation is usually formalized as $E(Y \mid X) = f(X, \beta)$. To carry out regression analysis, the form of the function f must be specified. Sometimes the form of this function is based on knowledge about the relationship between Y and X that does not rely on the data. If no such knowledge is available, a flexible or convenient form for f is chosen.

Assume now that the vector of unknown parameters β is of length k. In order to perform a regression analysis the user must provide information about the dependent variable Y:

- If N data points of the form (Y, X) are observed, where $N < k$, most classical approaches to regression analysis cannot be performed: since the system of equations defining the regression model is underdetermined, there are not enough data to recover β.

- If exactly $N = k$ data points are observed, and the function f is linear, the equations $Y = f(X, \beta)$ can be solved exactly rather than approximately. This reduces to solving a set of N equations with N unknowns (the elements of β), which has a unique solution as long as the X are linearly independent. If f is nonlinear, a solution may not exist, or many solutions may exist.

- The most common situation is where $N > k$ data points are observed. In this case, there is enough information in the data to estimate a unique value for β that best fits the data in some sense, and the regression model when applied to the data can be viewed as an overdetermined system in β.

In the last case, the regression analysis provides the tools for:

1. Finding a solution for unknown parameters β that will, for example, minimize the distance between the measured and predicted values of the dependent variable Y (also known as method of least squares).

2. Under certain statistical assumptions, the regression analysis uses the surplus of information to provide statistical information about the unknown parameters β and predicted values of the dependent variable Y.

Necessary Number of Independent Measurements

Consider a regression model which has three unknown parameters, β_0, β_1, and β_2. Suppose an experimenter performs 10 measurements all at exactly the same value of independent variable vector X (which contains the independent variables X_1, X_2, and X_3). In this case, regression analysis fails to give a unique set of estimated values for the three unknown parameters; the experimenter did not provide enough information. The best one can do is to estimate the average value and the standard deviation of the dependent variable Y. Similarly, measuring at two different values of X would give enough data for a regression with two unknowns, but not for three or more unknowns.

If the experimenter had performed measurements at three different values of the independent variable vector X, then regression analysis would provide a unique set of estimates for the three unknown parameters in β.

In the case of general linear regression, the above statement is equivalent to the requirement that the matrix X^TX is invertible.

Statistical Assumptions

When the number of measurements, N, is larger than the number of unknown parameters, k, and the measurement errors ϵ_i are normally distributed then *the excess of information* contained in $(N - k)$ measurements is used to make statistical predictions about the unknown parameters. This excess of information is referred to as the degrees of freedom of the regression.

Underlying Assumptions

Classical assumptions for regression analysis include:

- The sample is representative of the population for the inference prediction.

- The error is a random variable with a mean of zero conditional on the explanatory variables.

- The independent variables are measured with no error. (Note: If this is not so, modeling may be done instead using errors-in-variables model techniques).

- The independent variables (predictors) are linearly independent, i.e. it is not possible to express any predictor as a linear combination of the others.

- The errors are uncorrelated, that is, the variance–covariance matrix of the errors is diagonal and each non-zero element is the variance of the error.

- The variance of the error is constant across observations (homoscedasticity). If not, weighted least squares or other methods might instead be used.

These are sufficient conditions for the least-squares estimator to possess desirable properties; in particular, these assumptions imply that the parameter estimates will be unbiased, consistent, and efficient in the class of linear unbiased estimators. It is important to note that actual data rarely satisfies the assumptions. That is, the method is used even though the assumptions are not true. Variation from the assumptions can sometimes be used as a measure of how far the model is from being useful. Many of these assumptions may be relaxed in more advanced treatments. Reports of statistical analyses usually include analyses of tests on the sample data and methodology for the fit and usefulness of the model.

Assumptions include the geometrical support of the variables. Independent and dependent variables often refer to values measured at point locations. There may be spatial trends and spatial autocorrelation in the variables that violate statistical assumptions of regression. Geographic weighted regression is one technique to deal with such data. Also, variables may include values aggregated by areas. With aggregated data the modifiable areal unit problem can cause extreme variation in regression parameters. When analyzing data aggregated by political boundaries, postal codes or census areas results may be very distinct with a different choice of units.

Linear Regression

In linear regression, the model specification is that the dependent variable, y_i is a linear combination of the *parameters* (but need not be linear in the *independent variables*). For example, in simple linear regression for modeling n data points there is one independent variable: x_i, and two parameters, β_0 and β_1:

straight line: $y_i = \beta_0 + \beta_1 x_i + \varepsilon_i, \quad i = 1,\ldots,n.$

In multiple linear regression, there are several independent variables or functions of independent variables.

Adding a term in x_i^2 to the preceding regression gives:

parabola: $y_i = \beta_0 + \beta_1 x_i + \beta_2 x_i^2 + \varepsilon_i, i = 1,\ldots,n.$

This is still linear regression; although the expression on the right hand side is quadratic in the independent variable x_i, it is linear in the parameters β_0, β_1 and β_2.

In both cases, ε_i is an error term and the subscript i indexes a particular observation.

Returning our attention to the straight line case: Given a random sample from the population, we estimate the population parameters and obtain the sample linear regression model:

$$\widehat{y_i} = \hat{\beta}_0 + \hat{\beta}_1 x_i.$$

The residual, $e_i = y_i - \hat{y}_i$, is the difference between the value of the dependent variable predicted by the model, $\widehat{y_i}$, and the true value of the dependent variable, y_i. One method of estimation is ordinary least squares. This method obtains parameter estimates that minimize the sum of squared residuals, SSE, also sometimes denoted RSS:

$$SSE = \sum_{i=1}^{n} e_i^2.$$

Minimization of this function results in a set of normal equations, a set of simultaneous linear equations in the parameters, which are solved to yield the parameter estimators, $\hat{\beta}_0, \hat{\beta}_1$.

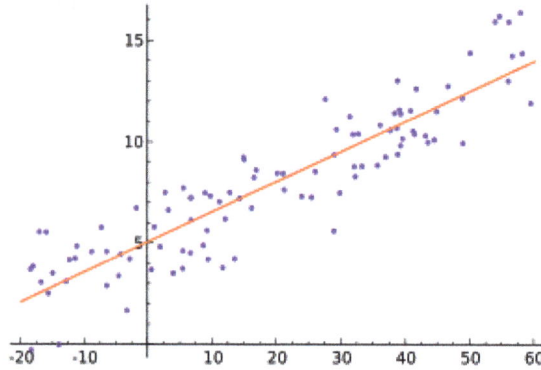

Illustration of linear regression on a data set.

In the case of simple regression, the formulas for the least squares estimates are

$$\widehat{\beta}_1 = \frac{\sum(x_i - \bar{x})(y_i - \bar{y})}{\sum(x_i - \bar{x})^2} \text{ and } \widehat{\beta}_0 = \bar{y} - \widehat{\beta}_1 \bar{x}$$

where $\bar{x}$ is the mean (average) of the x values and $\bar{y}$ is the mean of the y values.

Under the assumption that the population error term has a constant variance, the estimate of that variance is given by:

$$\hat{\sigma}_\varepsilon^2 = \frac{SSE}{n-2}.$$

This is called the mean square error (MSE) of the regression. The denominator is the sample size reduced by the number of model parameters estimated from the same data, $(n-p)$ for p regressors or $(n-p-1)$ if an intercept is used. In this case, $p=1$ so the denominator is n-2.

The standard errors of the parameter estimates are given by

$$\hat{\sigma}_{\beta_0} = \hat{\sigma}_\varepsilon \sqrt{\frac{1}{n} + \frac{\bar{x}^2}{\sum(x_i - \bar{x})^2}}$$

$$\hat{\sigma}_{\beta_1} = \hat{\sigma}_\varepsilon \sqrt{\frac{1}{\sum(x_i - \bar{x})^2}}.$$

Under the further assumption that the population error term is normally distributed, the researcher can use these estimated standard errors to create confidence intervals and conduct hypothesis tests about the population parameters.

General Linear Model

In the more general multiple regression model, there are p independent variables:

$$y_i = \beta_1 x_{i1} + \beta_2 x_{i2} + \cdots + \beta_p x_{ip} + \varepsilon_i,$$

where x_{ij} is the ith observation on the jth independent variable. If the first independent variable takes the value 1 for all i, $x_{i1} = 1$, then β_1 is called the regression intercept.

The least squares parameter estimates are obtained from p normal equations. The residual can be written as

$$\varepsilon_i = y_i - \hat{\beta}_1 x_{i1} - \cdots - \hat{\beta}_p x_{ip}.$$

The normal equations are

$$\sum_{i=1}^{n}\sum_{k=1}^{p} X_{ij} X_{ik} \hat{\beta}_k = \sum_{i=1}^{n} X_{ij} y_i, \, j = 1, \ldots, p.$$

In matrix notation, the normal equations are written as

$$(\mathbf{X}^\top \mathbf{X})\hat{\boldsymbol{\beta}} = \mathbf{X}^\top \mathbf{Y},$$

where the ij element of X is x_{ij}, the i element of the column vector Y is y_i, and the j element of $\hat{\beta}$ is $\hat{\beta}_j$. Thus X is $n \times p$, Y is $n \times 1$, and $\hat{\beta}$ is $p \times 1$. The solution is

$$\hat{\boldsymbol{\beta}} = (\mathbf{X}^\top \mathbf{X})^{-1} \mathbf{X}^\top \mathbf{Y}.$$

Diagnostics

Once a regression model has been constructed, it may be important to confirm the goodness of fit of the model and the statistical significance of the estimated parameters. Commonly used checks of goodness of fit include the R-squared, analyses of the pattern of residuals and hypothesis testing. Statistical significance can be checked by an F-test of the overall fit, followed by t-tests of individual parameters.

Interpretations of these diagnostic tests rest heavily on the model assumptions. Although examination of the residuals can be used to invalidate a model, the results of a t-test or F-test are sometimes more difficult to interpret if the model's assumptions are violated. For example, if the error term does not have a normal distribution, in small samples the estimated parameters will not follow normal distributions and complicate inference. With relatively large samples, however, a central limit theorem can be invoked such that hypothesis testing may proceed using asymptotic approximations.

Limited Dependent Variables

The phrase "limited dependent" is used in econometric statistics for categorical and constrained variables.

The response variable may be non-continuous ("limited" to lie on some subset of the real line). For binary (zero or one) variables, if analysis proceeds with least-squares linear regression, the model is called the linear probability model. Nonlinear models for binary dependent variables include the probit and logit model. The multivariate probit model is a standard method of estimating a joint relationship between several binary dependent variables and some independent variables. For categorical variables with more than two values there is the multinomial logit. For ordinal variables with more than two values, there are the ordered logit and ordered probit models. Censored regression models may be used when the dependent variable is only sometimes observed, and Heckman correction type models may be used when the sample is not randomly selected from the population of interest. An alternative to such procedures is linear regression based on polychoric correlation (or polyserial correlations) between the categorical variables. Such procedures differ in the assumptions made about the distribution of the variables in the population. If the variable is positive with low values and represents the repetition of the occurrence of an event, then count models like the Poisson regression or the negative binomial model may be used instead.

Interpolation and Extrapolation

Regression models predict a value of the Y variable given known values of the X variables. Prediction *within* the range of values in the dataset used for model-fitting is known informally as interpolation. Prediction *outside* this range of the data is known as extrapolation. Performing extrapolation relies strongly on the regression assumptions. The further the extrapolation goes outside the data, the more room there is for the model to fail due to differences between the assumptions and the sample data or the true values.

It is generally advised that when performing extrapolation, one should accompany the estimated value of the dependent variable with a prediction interval that represents the uncertainty. Such intervals tend to expand rapidly as the values of the independent variable(s) moved outside the range covered by the observed data.

For such reasons and others, some tend to say that it might be unwise to undertake extrapolation.

However, this does not cover the full set of modelling errors that may be being made: in particular, the assumption of a particular form for the relation between Y and X. A properly conducted regression analysis will include an assessment of how well the assumed form is matched by the observed data, but it can only do so within the range of values of the independent variables actually available. This means that any extrapolation is particularly reliant on the assumptions being made about the structural form of the regression relationship. Best-practice advice here is that a linear-in-variables and linear-in-parameters relationship should not be chosen simply for computational convenience, but that all available knowledge should be deployed in constructing a regression model. If this knowledge includes the fact that the dependent variable cannot go outside a certain range of values, this can be made use of in selecting the model – even if the observed dataset has no values particularly near such bounds. The implications of this step of choosing an appropriate functional form for the regression can be great when extrapolation is considered. At a minimum, it can ensure that any extrapolation arising from a fitted model is "realistic" (or in accord with what is known).

Nonlinear Regression

When the model function is not linear in the parameters, the sum of squares must be minimized by an iterative procedure. This introduces many complications which are summarized in differences between linear and non-linear least squares.

Power and Sample Size Calculations

There are no generally agreed methods for relating the number of observations versus the number of independent variables in the model. One rule of thumb suggested by Good and Hardin is $N = m^n$, where N is the sample size, n is the number of independent variables and m is the number of observations needed to reach the desired precision if the model had only one independent variable. For example, a researcher is building a linear regression model using a dataset that contains 1000 patients (N). If the researcher decides that five observations are needed to precisely define a straight line (m), then the maximum number of independent variables the model can support is 4, because

$$\frac{\log 1000}{\log 5} = 4.29.$$

Other Methods

Although the parameters of a regression model are usually estimated using the method of least squares, other methods which have been used include:

- Bayesian methods, e.g. Bayesian linear regression

- Percentage regression, for situations where reducing *percentage* errors is deemed more appropriate.

- Least absolute deviations, which is more robust in the presence of outliers, leading to quantile regression

- Nonparametric regression, requires a large number of observations and is computationally intensive

- Distance metric learning, which is learned by the search of a meaningful distance metric in a given input space.

Software

All major statistical software packages perform least squares regression analysis and inference. Simple linear regression and multiple regression using least squares can be done in some spreadsheet applications and on some calculators. While many statistical software packages can perform various types of nonparametric and robust regression, these methods are less standardized; different software packages implement different methods, and a method with a given name may be implemented differently in different packages. Specialized regression software has been developed for use in fields such as survey analysis and neuroimaging.

Linear Regression

In statistics, linear regression is an approach for modeling the relationship between a scalar dependent variable y and one or more explanatory variables (or independent variables) denoted X. The case of one explanatory variable is called *simple linear regression*. For more than one explanatory variable, the process is called *multiple linear regression*. (This term is distinct from *multivariate linear regression*, where multiple correlated dependent variables are predicted, rather than a single scalar variable.)

In linear regression, the relationships are modeled using linear predictor functions whose unknown model parameters are estimated from the data. Such models are called *linear models*. Most commonly, the conditional mean of y given the value of X is assumed to be an affine function of X; less commonly, the median or some other quantile of the conditional distribution of y given X is expressed as a linear function of X. Like all forms of regression analysis, linear regression focuses on the conditional probability distribution of y given X, rather than on the joint probability distribution of y and X, which is the domain of multivariate analysis.

Linear regression was the first type of regression analysis to be studied rigorously, and to be used extensively in practical applications. This is because models which depend linearly on their unknown parameters are easier to fit than models which are non-linearly related to their parameters and because the statistical properties of the resulting estimators are easier to determine.

Linear regression has many practical uses. Most applications fall into one of the following two broad categories:

- If the goal is prediction, or forecasting, or error reduction, linear regression can be used to fit a predictive model to an observed data set of y and X values. After developing such a model, if an additional value of X is then given without its accompanying value of y, the fitted model can be used to make a prediction of the value of y.

- Given a variable y and a number of variables X_1, ..., X_p that may be related to y, linear regression analysis can be applied to quantify the strength of the relationship between y and the X_j, to assess which X_j may have no relationship with y at all, and to identify which subsets of the X_j contain redundant information about y.

Linear regression models are often fitted using the least squares approach, but they may also be fitted in other ways, such as by minimizing the "lack of fit" in some other norm (as with least absolute deviations regression), or by minimizing a penalized version of the least squares loss function as in ridge regression (L^2-norm penalty) and lasso (L^1-norm penalty). Conversely, the least squares approach can be used to fit models that are not linear models. Thus, although the terms "least squares" and "linear model" are closely linked, they are not synonymous.

Introduction

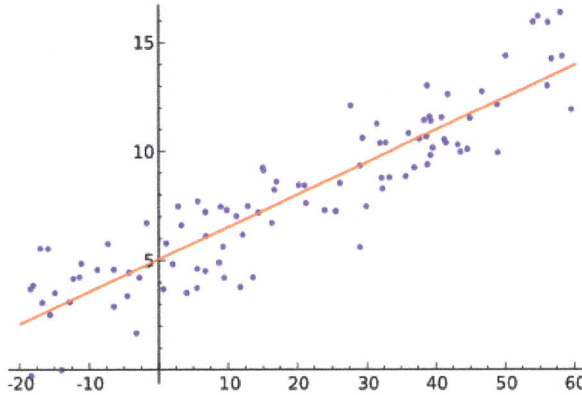

Example of simple linear regression, which has one independent variable

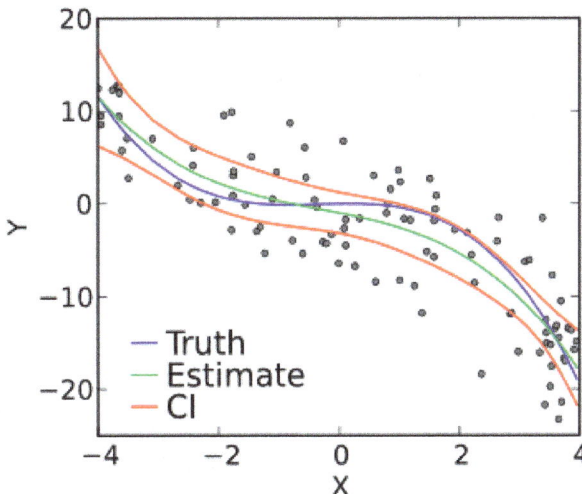

Example of a cubic polynomial regression, which is a type of linear regression.

Given a data set $\{y_i, x_{i1}, \ldots, x_{ip}\}_{i=1}^{n}$ of n statistical units, a linear regression model assumes that the relationship between the dependent variable y_i and the p-vector of regressors x_i is linear. This relationship is modeled through a *disturbance term* or *error variable* ε_i — an unobserved random variable that adds noise to the linear relationship between the dependent variable and regressors. Thus the model takes the form

$$y_i = \beta_0 1 + \beta_1 x_{i1} + \cdots + \beta_p x_{ip} + \varepsilon_i = x_i^\top \beta + \varepsilon_i, \qquad i = 1, \ldots, n,$$

where $^\top$ denotes the transpose, so that $x_i^\top \beta$ is the inner product between vectors x_i and β.

Often these n equations are stacked together and written in vector form as

$$\mathbf{y} = X\boldsymbol{\beta} + \boldsymbol{\varepsilon},$$

where

$$\mathbf{y} = \begin{pmatrix} y_1 \\ y_2 \\ \vdots \\ y_n \end{pmatrix},$$

$$X = \begin{pmatrix} \mathbf{x}_1^\top \\ \mathbf{x}_2^\top \\ \vdots \\ \mathbf{x}_n^\top \end{pmatrix} = \begin{pmatrix} 1 & x_{11} & \cdots & x_{1p} \\ 1 & x_{21} & \cdots & x_{2p} \\ \vdots & \vdots & \ddots & \vdots \\ 1 & x_{n1} & \cdots & x_{np} \end{pmatrix},$$

$$\boldsymbol{\beta} = \begin{pmatrix} \beta_0 \\ \beta_1 \\ \beta_2 \\ \vdots \\ \beta_p \end{pmatrix}, \quad \boldsymbol{\varepsilon} = \begin{pmatrix} \varepsilon_1 \\ \varepsilon_2 \\ \vdots \\ \varepsilon_n \end{pmatrix}.$$

Some remarks on terminology and general use:

- y_i is called the *regressand, endogenous variable, response variable, measured variable, criterion variable,* or *dependent variable*. The decision as to which variable in a data set is modeled as the dependent variable and which are modeled as the independent variables may be based on a presumption that the value of one of the variables is caused by, or directly influenced by the other variables. Alternatively, there may be an operational reason to model one of the variables in terms of the others, in which case there need be no presumption of causality.

- $x_{i1}, x_{i2}, \ldots, x_{ip}$ are called *regressors, exogenous variables, explanatory variables, covariates, input variables, predictor variables,* or *independent variables.* The matrix X is sometimes called the design matrix.

  o Usually a constant is included as one of the regressors. For example, we can take $x_{i1} = 1$ for $i = 1, \ldots, n$. The corresponding element of β is called the *intercept.* Many statistical inference procedures for linear models require an intercept to be present, so it is often included even if theoretical considerations suggest that its value should be zero.

  o Sometimes one of the regressors can be a non-linear function of another regressor or of the data, as in polynomial regression and segmented regression. The model remains linear as long as it is linear in the parameter vector β.

  o The regressors x_{ij} may be viewed either as random variables, which we simply observe, or they can be considered as predetermined fixed values which we can choose. Both interpretations may be appropriate in different cases, and they generally lead to the same estimation procedures; however different approaches to asymptotic analysis are used in these two situations.

- β is a *(p+1)*-dimensional *parameter vector.* Where β_0 is the constant (offset) term. In figure Simple linear regression with one variable $\beta_0 = 5$. Its elements are also called *effects,* or *regression coefficients.* Statistical estimation and inference in linear regression focuses on β. The elements of this parameter vector are interpreted as the partial derivatives of the dependent variable with respect to the various independent variables.

- ε_i is called the *error term, disturbance term,* or *noise.* This variable captures all other factors which influence the dependent variable y_i other than the regressors x_i. The relationship between the error term and the regressors, for example whether they are correlated, is a crucial step in formulating a linear regression model, as it will determine the method to use for estimation.

Example. Consider a situation where a small ball is being tossed up in the air and then we measure its heights of ascent h_i at various moments in time t_i. Physics tells us that, ignoring the drag, the relationship can be modeled as

$$h_i = \beta_1 t_i + \beta_2 t_i^2 + \varepsilon_i,$$

where β_1 determines the initial velocity of the ball, β_2 is proportional to the standard gravity, and ε_i is due to measurement errors. Linear regression can be used to estimate the values of β_1 and β_2 from the measured data. This model is non-linear in the time variable, but it is linear in the parameters β_1 and β_2; if we take regressors $x_i = (x_{i1}, x_{i2}) = (t_i, t_i^2)$, the model takes on the standard form

$$h_i = \mathrm{x}_i^\top \beta + \varepsilon_i.$$

Assumptions

Standard linear regression models with standard estimation techniques make a number of assumptions about the predictor variables, the response variables and their relationship. Numerous extensions have been developed that allow each of these assumptions to be relaxed (i.e. reduced to a weaker form), and in some cases eliminated entirely. Some methods are general enough that they can relax multiple assumptions at once, and in other cases this can be achieved by combining different extensions. Generally these extensions make the estimation procedure more complex and time-consuming, and may also require more data in order to produce an equally precise model.

The following are the major assumptions made by standard linear regression models with standard estimation techniques (e.g. ordinary least squares):

- Weak exogeneity. This essentially means that the predictor variables x can be treated as fixed values, rather than random variables. This means, for example, that the predictor variables are assumed to be error-free—that is, not contaminated with measurement errors. Although this assumption is not realistic in many settings, dropping it leads to significantly more difficult errors-in-variables models.

- Linearity. This means that the mean of the response variable is a linear combination of the parameters (regression coefficients) and the predictor variables. Note that this assumption is much less restrictive than it may at first seem. Because the predictor variables are treated as fixed values, linearity is really only a restriction on the parameters. The predictor variables themselves can be arbitrarily transformed, and in fact multiple copies of the same underlying predictor variable can be added, each one transformed differently. This trick is used, for example, in polynomial regression, which uses linear regression to fit the response variable as an arbitrary polynomial function (up to a given rank) of a predictor variable. This makes linear regression an extremely powerful inference method. In fact, models such as polynomial regression are often "too powerful", in that they tend to overfit the data. As a result, some kind of regularization must typically be used to prevent unreasonable solutions coming out of the estimation process. Common examples are ridge regression and lasso regression. Bayesian linear regression can also be used, which by its nature is more or less immune to the problem of overfitting. (In fact, ridge regression and lasso regression can both be viewed as special cases of Bayesian linear regression, with particular types of prior distributions placed on the regression coefficients.)

- Constant variance (a.k.a. homoscedasticity). This means that different response variables have the same variance in their errors, regardless of the values of the predictor variables. In practice this assumption is invalid (i.e. the errors are heteroscedastic) if the response variables can vary over a wide scale. In order to determine for heterogeneous error variance, or when a pattern of residuals violates model assumptions of homoscedasticity (error is equally variable around the 'best-fitting line' for all points of x), it is prudent to look for a "fanning effect" between residual error and predicted values. This is to say there will be a systematic change in the absolute or squared residuals when plotted against the predicting outcome. Error will not be evenly distributed across the regression line. Heteroscedasticity will result in the averaging over of distinguishable variances around the points to get a single variance that is inaccurately representing all the variances of the line. In effect, residuals appear clustered and spread apart on their predicted plots for larger and smaller values for points along the linear regression line, and the mean squared error for the model will be wrong. Typically, for example, a response variable whose mean is large will have a greater variance than one whose mean is small. For example, a given person whose income is predicted to be $100,000 may easily have an actual income of $80,000 or $120,000 (a standard deviation of around $20,000), while another person with a predicted income of $10,000 is unlikely to have the same $20,000 standard deviation, which would imply their actual income would vary anywhere between -$10,000 and $30,000. (In fact, as this shows, in many cases—often the same cases where the assumption of normally distributed errors fails—the variance or standard deviation should be predicted to be proportional to the mean, rather than constant.) Simple linear regression estimation methods give less precise parameter estimates and misleading inferential quantities such as standard errors when substantial heteroscedasticity is present. However, various estimation techniques (e.g. weighted least squares and heteroscedasticity-consistent standard errors) can handle heteroscedasticity in a quite general way. Bayesian linear regression techniques can also be used when the variance is assumed to be a function of the mean. It is also possible in some cases to fix the problem by applying a transformation to the response variable (e.g. fit the logarithm of the response variable using a linear regression model, which implies that the response variable has a log-normal distribution rather than a normal distribution).

- Independence of errors. This assumes that the errors of the response variables are uncorrelated with each other. (Actual statistical independence is a stronger condition than mere lack of correlation and is often not needed, although it can be exploited if it is known to hold.) Some methods (e.g. generalized least squares) are capable of handling correlated errors, although they typically require significantly more data unless some sort of regularization is used to bias the model towards assuming uncorrelated errors. Bayesian linear regression is a general way of handling this issue.

- Lack of multicollinearity in the predictors. For standard least squares estimation methods, the design matrix X must have full column rank p; otherwise, we have a condition known as multicollinearity in the predictor variables. This can be triggered by having two or more perfectly correlated predictor variables (e.g. if the same predictor variable is mistakenly given twice, either without transforming one of the copies or by transforming one of the copies linearly). It can also happen if there is too little data available compared to the number of parameters to be estimated (e.g. fewer data points than regression coefficients). In the case of multicollinearity, the parameter vector β will be non-identifiable—it has no unique solution. At most we will be able to identify some of the parameters, i.e. narrow down its value to some linear subspace of R^p. Methods for fitting linear models with multicollinearity have been developed; some require additional assumptions such as "effect sparsity"—that a large fraction of the effects are exactly zero.

Note that the more computationally expensive iterated algorithms for parameter estimation, such as those used in generalized linear models, do not suffer from this problem—and in fact it's quite normal when handling categorically valued predictors to introduce a separate indicator variable predictor for each possible category, which inevitably introduces multicollinearity.

Beyond these assumptions, several other statistical properties of the data strongly influence the performance of different estimation methods:

- The statistical relationship between the error terms and the regressors plays an important role in determining whether an estimation procedure has desirable sampling properties such as being unbiased and consistent.

- The arrangement, or probability distribution of the predictor variables x has a major influence on the precision of estimates of β. Sampling and design of experiments are highly developed subfields of statistics that provide guidance for collecting data in such a way to achieve a precise estimate of β.

Interpretation

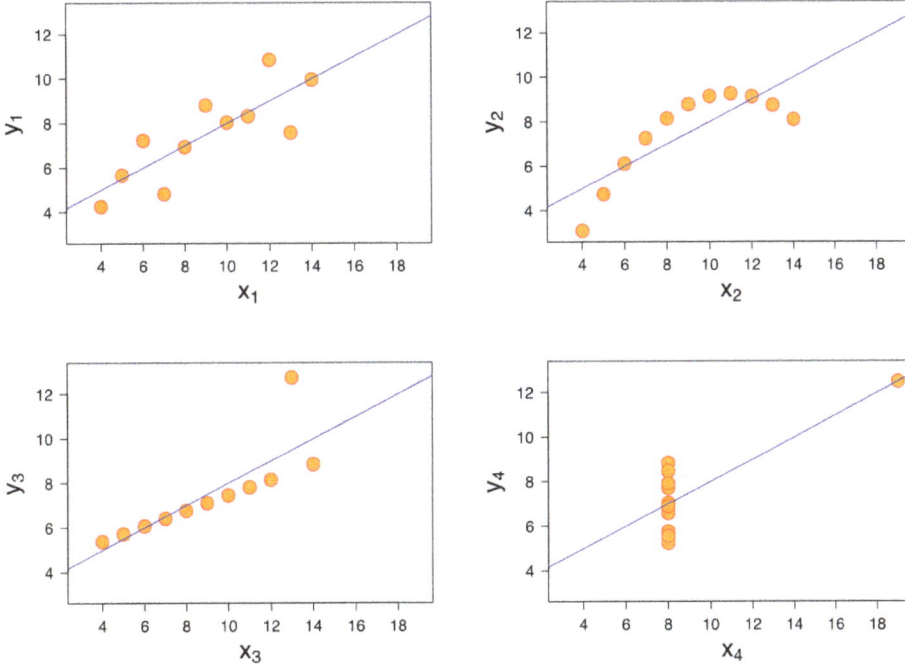

The data sets in the Anscombe's quartet are designed to have the same linear regression line but are graphically very different.

A fitted linear regression model can be used to identify the relationship between a single predictor variable x_j and the response variable y when all the other predictor variables in the model are "held fixed". Specifically, the interpretation of β_j is the expected change in y for a one-unit change in x_j when the other covariates are held fixed—that is, the expected value of the partial derivative of y with respect to x_j. This is sometimes called the *unique effect* of x_j on y. In contrast, the *marginal effect* of x_j on y can be assessed using a correlation coefficient or simple linear regression model relating x_j to y; this effect is the total derivative of y with respect to x_j.

Care must be taken when interpreting regression results, as some of the regressors may not allow for marginal changes (such as dummy variables, or the intercept term), while others cannot be held fixed (recall the example from the introduction: it would be impossible to "hold t_i fixed" and at the same time change the value of t_i^2).

It is possible that the unique effect can be nearly zero even when the marginal effect is large. This may imply that some other covariate captures all the information in x_j, so that once that variable is in the model, there is no contribution of x_j to the variation in y. Conversely, the unique effect of x_j can be large while its marginal effect is nearly zero. This would happen if the other covariates explained a great deal of the variation of y, but they mainly explain variation in a way that is complementary to what is captured by x_j. In this case, including the other variables in the model reduces the part of the variability of y that is unrelated to x_j, thereby strengthening the apparent relationship with x_j.

The meaning of the expression "held fixed" may depend on how the values of the predictor variables arise. If the experimenter directly sets the values of the predictor variables according to a study design, the comparisons of interest may literally correspond to comparisons among units whose predictor variables have been "held fixed" by the experimenter. Alternatively, the expression "held fixed" can refer to a selection that takes place in the context of data analysis. In this case, we "hold a variable fixed" by restricting our attention to the subsets of the data that happen to have a common value for the given predictor variable. This is the only interpretation of "held fixed" that can be used in an observational study.

The notion of a "unique effect" is appealing when studying a complex system where multiple interrelated components influence the response variable. In some cases, it can literally be interpreted as the causal effect of an intervention that is linked to the value of a predictor variable. However, it has been argued that in many cases multiple regression analysis fails to clarify the relationships between the predictor variables and the response variable when the predictors are correlated with each other and are not assigned following a study design. A commonality analysis may be helpful in disentangling the shared and unique impacts of correlated independent variables.

Extensions

Numerous extensions of linear regression have been developed, which allow some or all of the assumptions underlying the basic model to be relaxed.

Simple and Multiple Regression

The very simplest case of a single scalar predictor variable x and a single scalar response variable y is known as *simple linear regression*. The extension to multiple and/or vector-valued predictor variables (denoted with a capital X) is known as *multiple linear regression*, also known as *multivariable linear regression*. Nearly all real-world regression models involve multiple predictors, and basic descriptions of linear regression are often phrased in terms of the multiple regression model. Note, however, that in these cases the response variable y is still a scalar. Another term *multivariate linear regression* refers to cases where y is a vector, i.e., the same as *general linear regression*.

General Linear Models

The general linear model considers the situation when the response variable Y is not a scalar but a vector. Conditional linearity of $E(y|x) = Bx$ is still assumed, with a matrix B replacing the vector β of the classical linear regression model. Multivariate analogues of Ordinary Least-Squares (OLS) and Generalized Least-Squares (GLS) have been developed. "General linear models" are also called "multivariate linear models". These are not the same as multivariable linear models (also called "multiple linear models").

Heteroscedastic Models

Various models have been created that allow for heteroscedasticity, i.e. the errors for different response variables may have different variances. For example, weighted least squares is a method for estimating linear regression models when the response variables may have different error variances, possibly with correlated errors. Heteroscedasticity-consistent standard errors is an improved method for use with uncorrelated but potentially heteroscedastic errors.

Generalized Linear Models

Generalized linear models (GLMs) are a framework for modeling a response variable y that is bounded or discrete. This is used, for example:

- when modeling positive quantities (e.g. prices or populations) that vary over a large scale—which are better described using a skewed distribution such as the log-normal distribution or Poisson distribution (although GLMs are not used for log-normal data, instead the response variable is simply transformed using the logarithm function);

- when modeling categorical data, such as the choice of a given candidate in an election (which is better described using a Bernoulli distribution/binomial distribution for binary choices, or a categorical distribution/multinomial distribution for multi-way choices), where there are a fixed number of choices that cannot be meaningfully ordered;

- when modeling ordinal data, e.g. ratings on a scale from 0 to 5, where the different outcomes can be ordered but where the quantity itself may not have any absolute meaning (e.g. a rating of 4 may not be "twice as good" in any objective sense as a rating of 2, but simply indicates that it is better than 2 or 3 but not as good as 5).

Generalized linear models allow for an arbitrary *link function g* that relates the mean of the response variable to the predictors, i.e. $E(y) = g(\beta'x)$. The link function is often related to the distribution of the response, and in particular it typically has the effect of transforming between the $(-\infty, \infty)$ range of the linear predictor and the range of the response variable.

Some common examples of GLMs are:

- Poisson regression for count data.

- Logistic regression and probit regression for binary data.

- Multinomial logistic regression and multinomial probit regression for categorical data.

- Ordered probit regression for ordinal data.

Single index models allow some degree of nonlinearity in the relationship between x and y, while preserving the central role of the linear predictor $\beta'x$ as in the classical linear regression model. Under certain conditions, simply applying OLS to data from a single-index model will consistently estimate β up to a proportionality constant.

Hierarchical Linear Models

Hierarchical linear models (or *multilevel regression*) organizes the data into a hierarchy of regressions, for example where A is regressed on B, and B is regressed on C. It is often used where the variables of interest have a natural hierarchical structure such as in educational statistics, where students are nested in classrooms, classrooms are nested in schools, and schools are nested in some administrative grouping, such as a school district. The response variable might be a measure of student achievement such as a test score, and different co-variates would be collected at the classroom, school, and school district levels.

Errors-in-variables

Errors-in-variables models (or "measurement error models") extend the traditional linear regression model to allow the predictor variables X to be observed with error. This error causes standard estimators of β to become biased. Generally, the form of bias is an attenuation, meaning that the effects are biased toward zero.

Others

- In Dempster–Shafer theory, or a linear belief function in particular, a linear regression model may be represented as a partially swept matrix, which can be combined with similar matrices representing observations and other assumed normal distributions and state equations. The combination of swept or unswept matrices provides an alternative method for estimating linear regression models.

Estimation Methods

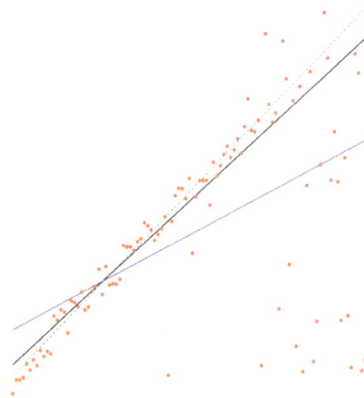

Comparison of the Theil–Sen estimator (black) and simple linear regression (blue) for a set of points with outliers.

A large number of procedures have been developed for parameter estimation and inference in linear regression. These methods differ in computational simplicity of algorithms, presence of a closed-form solution, robustness with respect to heavy-tailed distributions, and theoretical assumptions needed to validate desirable statistical properties such as consistency and asymptotic efficiency.

Some of the more common estimation techniques for linear regression are summarized below.

Least-squares Estimation and Related Techniques

- Ordinary least squares (OLS) is the simplest and thus most common estimator. It is conceptually simple and computationally straightforward. OLS estimates are commonly used to analyze both experimental and observational data.

  The OLS method minimizes the sum of squared residuals, and leads to a closed-form expression for the estimated value of the unknown parameter β:

$$\hat{\beta} = (\mathbf{X}^\top \mathbf{X})^{-1} \mathbf{X}^\top \mathbf{y} = \left(\sum \mathbf{x}_i \mathbf{x}_i^\top \right)^{-1} \left(\sum \mathbf{x}_i y_i \right).$$

  The estimator is unbiased and consistent if the errors have finite variance and are uncorrelated with the regressors

$$\mathrm{E}[\mathbf{x}_i \varepsilon_i] = 0.$$

  It is also efficient under the assumption that the errors have finite variance and are homoscedastic, meaning that $\mathrm{E}[\varepsilon_i^2 | \mathbf{x}_i]$ does not depend on i. The condition that the errors are uncorrelated with the regressors will generally be satisfied in an experiment, but in the case of observational data, it is difficult to exclude the possibility of an omitted covariate z that is related to both the observed covariates and the response variable. The existence of such a covariate will generally lead to a correlation between the regressors and the response variable, and hence to an inconsistent estimator of β. The condition of homoscedasticity can fail with either experimental or observational data. If the goal is either inference or predictive modeling, the performance of OLS estimates can be poor if multicollinearity is present, unless the sample size is large.

  In simple linear regression, where there is only one regressor (with a constant), the OLS coefficient estimates have a simple form that is closely related to the correlation coefficient between the covariate and the response.

- Generalized least squares (GLS) is an extension of the OLS method, that allows efficient estimation of β when either heteroscedasticity, or correlations, or both are present among the error terms of the model, as long as the form of heteroscedasticity and correlation is known independently of the data. To handle heteroscedasticity when the error terms are uncorrelated with each other, GLS minimizes a weighted analogue to the sum of squared residuals from OLS regression, where the weight for the ith case is inversely proportional to $\text{var}(\varepsilon_i)$. This special case of GLS is called "weighted least squares". The GLS solution to estimation problem is

$$\hat{\beta} = (\mathbf{X}^\top \Omega^{-1} \mathbf{X})^{-1} \mathbf{X}^\top \Omega^{-1} \mathbf{y},$$

where Ω is the covariance matrix of the errors. GLS can be viewed as applying a linear transformation to the data so that the assumptions of OLS are met for the transformed data. For GLS to be applied, the covariance structure of the errors must be known up to a multiplicative constant.

- Percentage least squares focuses on reducing percentage errors, which is useful in the field of forecasting or time series analysis. It is also useful in situations where the dependent variable has a wide range without constant variance, as here the larger residuals at the upper end of the range would dominate if OLS were used. When the percentage or relative error is normally distributed, least squares percentage regression provides maximum likelihood estimates. Percentage regression is linked to a multiplicative error model, whereas OLS is linked to models containing an additive error term.

- Iteratively reweighted least squares (IRLS) is used when heteroscedasticity, or correlations, or both are present among the error terms of the model, but where little is known about the covariance structure of the errors independently of the data. In the first iteration, OLS, or GLS with a provisional covariance structure is carried out, and the residuals are obtained from the fit. Based on the residuals, an improved estimate of the covariance structure of the errors can usually be obtained. A subsequent GLS iteration is then performed using this estimate of the error structure to define the weights. The process can be iterated to convergence, but in many cases, only one iteration is sufficient to achieve an efficient estimate of β.

- Instrumental variables regression (IV) can be performed when the regressors are correlated with the errors. In this case, we need the existence of some auxiliary *instrumental variables* z_i such that $E[z_i \varepsilon_i] = 0$. If Z is the matrix of instruments, then the estimator can be given in closed form as

$$\hat{\beta} = (\mathbf{X}^\top \mathbf{Z} (\mathbf{Z}^\top \mathbf{Z})^{-1} \mathbf{Z}^\top \mathbf{X})^{-1} \mathbf{X}^\top \mathbf{Z} (\mathbf{Z}^\top \mathbf{Z})^{-1} \mathbf{Z}^\top \mathbf{y}.$$

- Optimal instruments regression is an extension of classical IV regression to the situation where $E[\varepsilon_i \mid z_i] = 0$.

- Total least squares (TLS) is an approach to least squares estimation of the linear regression model that treats the covariates and response variable in a more geometrically symmetric manner than OLS. It is one approach to handling the "errors in variables" problem, and is also sometimes used even when the covariates are assumed to be error-free.

Maximum-likelihood Estimation and Related Techniques

- Maximum likelihood estimation can be performed when the distribution of the error terms is known to belong to a certain parametric family f_θ of probability distributions. When f_θ is a normal distribution with zero mean and variance θ, the resulting estimate is identical to the OLS estimate. GLS estimates are maximum likelihood estimates when ε follows a multivariate normal distribution with a known covariance matrix.

- Ridge regression, and other forms of penalized estimation such as Lasso regression, deliberately introduce bias into the estimation of β in order to reduce the variability of the estimate. The resulting estimators generally have lower mean squared error than the OLS estimates, particularly when multicollinearity is present or when overfitting is a problem. They are generally used when the goal is to predict the value of the response variable y for values of the predictors x that have not yet been observed. These methods are not as commonly used when the goal is inference, since it is difficult to account for the bias.

- Least absolute deviation (LAD) regression is a robust estimation technique in that it is less sensitive to the presence of outliers than OLS (but is less efficient than OLS when no outliers are present). It is equivalent to maximum likelihood estimation under a Laplace distribution model for ε.

- Adaptive estimation. If we assume that error terms are independent from the regressors, the optimal estimator is the 2-step MLE, where the first step is used to non-parametrically estimate the distribution of the error term.

Other Estimation Techniques

- Bayesian linear regression applies the framework of Bayesian statistics to linear regression. In particular, the regression coefficients β are assumed to be random variables with a specified prior distribution. The prior distribution can bias the solutions for the regression coefficients, in a way similar to (but more general than) ridge regression or lasso regression. In addition, the Bayesian estimation process produces not a single point estimate for the "best" values of the regression coefficients but an entire posterior distribution, completely describing the uncertainty surrounding the quantity. This can be used to estimate the "best" coefficients using the mean, mode, median, any quantile or any other function of the posterior distribution.

- Quantile regression focuses on the conditional quantiles of y given X rather than the conditional mean of y given X. Linear quantile regression models a particular conditional quantile, for example the conditional median, as a linear function $\beta^{\mathrm{T}}x$ of the predictors.

- Mixed models are widely used to analyze linear regression relationships involving dependent data when the dependencies have a known structure. Common applications of mixed models include analysis of data involving repeated measurements, such as longitudinal data, or data obtained from cluster sampling. They are generally fit as parametric models, using maximum likelihood or Bayesian estimation. In the case where the errors are modeled as normal random variables, there is a close connection between mixed models and generalized least squares. Fixed effects estimation is an alternative approach to analyzing this type of data.

- Principal component regression (PCR) is used when the number of predictor variables is large, or when strong correlations exist among the predictor variables. This two-stage procedure first reduces the predictor variables using principal component analysis then uses the reduced variables in an OLS regression fit. While it often works well in practice, there is no general theoretical reason that the most informative linear function of the predictor variables should lie among the dominant principal components of the multivariate distribution of the predictor variables. The partial least squares regression is the extension of the PCR method which does not suffer from the mentioned deficiency.

- Least-angle regression is an estimation procedure for linear regression models that was developed to handle high-dimensional covariate vectors, potentially with more covariates than observations.

- The Theil–Sen estimator is a simple robust estimation technique that chooses the slope of the fit line to be the median of the slopes of the lines through pairs of sample points. It has similar statistical efficiency properties to simple linear regression but is much less sensitive to outliers.

- Other robust estimation techniques, including the α-trimmed mean approach, and L-, M-, S-, and R-estimators have been introduced.

Further Discussion

In statistics and numerical analysis, the problem of numerical methods for linear least squares is an important one because linear regression models are one of the most important types of model, both as formal statistical models and for exploration of data sets. The majority of statistical computer packages contain facilities for regression analysis that make use of linear least squares computations. Hence it is appropriate that considerable effort has been devoted to the task of ensuring that these computations are undertaken efficiently and with due regard to numerical precision.

Individual statistical analyses are seldom undertaken in isolation, but rather are part of a sequence of investigatory steps. Some of the topics involved in considering numerical methods for linear least squares relate to this point. Thus important topics can be

- Computations where a number of similar, and often nested, models are considered for the same data set. That is, where models with the same dependent variable but different sets of independent variables are to be considered, for essentially the same set of data points.

- Computations for analyses that occur in a sequence, as the number of data points increases.

- Special considerations for very extensive data sets.

Fitting of linear models by least squares often, but not always, arises in the context of statistical analysis. It can therefore be important that considerations of computational efficiency for such problems extend to all of the auxiliary quantities required for such analyses, and are not restricted to the formal solution of the linear least squares problem.

Matrix calculations, like any others, are affected by rounding errors. An early summary of these effects, regarding the choice of computational methods for matrix inversion, was provided by Wilkinson.

Using Linear Algebra

It follows that one can find a "best" approximation of another function by minimizing the area between two functions, a continuous function f on $[a,b]$ and a function $g \in W$ where W is a subspace of $C[a,b]$:

$$Area = \int_a^b |f(x) - g(x)| dx,$$

all within the subspace W. Due to the frequent difficulty of evaluating integrands involving absolute value, one can instead define

$$\int_a^b [f(x) - g(x)]^2 dx$$

as an adequate criterion for obtaining the least squares approximation, function g, of f with respect to the inner product space W.

As such, $\|f-g\|^2$ or, equivalently, $\|f-g\|$, can thus be written in vector form:

$$\int_a^b [f(x)-g(x)]^2\,dx = \langle f-g, f-g\rangle = \|f-g\|^2.$$

In other words, the least squares approximation of f is the function $g \in$ *subspace W* closest to f in terms of the inner product $\langle f,g\rangle$. Furthermore, this can be applied with a theorem:

Let f be continuous on $[a,b]$, , and let W be a finite-dimensional subspace of $C[a,b]$. The least squares approximating function of f with respect to W is given by

$$g = \langle f, \vec{w_1}\rangle \vec{w_1} + \langle f, \vec{w_2}\rangle \vec{w_2} + \cdots + \langle f, \vec{w_n}\rangle \vec{w_n},$$

where $B = \{\vec{w_1}, \vec{w_2}, \ldots, \vec{w_n}\}$ is an orthonormal basis for W.

Applications of Linear Regression

Linear regression is widely used in biological, behavioral and social sciences to describe possible relationships between variables. It ranks as one of the most important tools used in these disciplines.

Trend Line

A trend line represents a trend, the long-term movement in time series data after other components have been accounted for. It tells whether a particular data set (say GDP, oil prices or stock prices) have increased or decreased over the period of time. A trend line could simply be drawn by eye through a set of data points, but more properly their position and slope is calculated using statistical techniques like linear regression. Trend lines typically are straight lines, although some variations use higher degree polynomials depending on the degree of curvature desired in the line.

Trend lines are sometimes used in business analytics to show changes in data over time. This has the advantage of being simple. Trend lines are often used to argue that a particular action or event (such as training, or an advertising campaign) caused observed changes at a point in time. This is a simple technique, and does not require a control group, experimental design, or a sophisticated analysis technique. However, it suffers from a lack of scientific validity in cases where other potential changes can affect the data.

Epidemiology

Early evidence relating tobacco smoking to mortality and morbidity came from observational studies employing regression analysis. In order to reduce spurious correlations when analyzing observational data, researchers usually include several variables in their regression models in addition to the variable of primary interest. For example, suppose we have a regression model in which cigarette smoking is the independent variable of interest, and the dependent variable is lifespan measured in years. Researchers might include socio-economic status as an additional independent variable, to ensure that any observed effect of smoking on lifespan is not due to some effect of education or income. However, it is never possible to include all possible confounding variables in an empirical analysis. For example, a hypothetical gene might increase mortality and also cause people to smoke more. For this reason, randomized controlled trials are often able to generate more compelling evidence of causal relationships than can be obtained using regression analyses of observational data. When controlled experiments are not feasible, variants of regression analysis such as instrumental variables regression may be used to attempt to estimate causal relationships from observational data.

Finance

The capital asset pricing model uses linear regression as well as the concept of beta for analyzing and quantifying the systematic risk of an investment. This comes directly from the beta coefficient of the linear regression model that relates the return on the investment to the return on all risky assets.

Economics

Linear regression is the predominant empirical tool in economics. For example, it is used to predict consumption spending, fixed investment spending, inventory investment, purchases of a country's exports, spending on imports, the demand to hold liquid assets, labor demand, and labor supply.

Environmental Science

Linear regression finds application in a wide range of environmental science applications. In Canada, the Environmental Effects Monitoring Program uses statistical analyses on fish and benthic surveys to measure the effects of pulp mill or metal mine effluent on the aquatic ecosystem.

Estimation of Multiple Regression Coefficients

In multiple linear regression, we essentially solve n equations for the p unknown parameters. Thus n must be equal to or greater than p and in practice n should be at least 3 or 4 times as large as p. The difference between the observed and predicted value of y (using regression) or the error is $= y_i - \hat{y}_i$ The regression coefficients are obtained by minimizing the sum of squares of errors.

In matrix form, the n equations can be written as

$$Y = Xb + e$$

where $Y = (n \times 1)$ column vector of the dependent variable, $X = (n \times p)$ matrix of independent variables, $b = (p \times 1)$ column vector of the regression coefficients, and $e = (n \times 1)$ column vector of residuals. The residuals are conditioned by:

$$E[e] = 0$$

$$Cov(e) = \sigma_\varepsilon^2 I$$

where $I = (n \times n)$ diagonal identity matrix with diagonal elements $= 1$ and off-diagonal elements $= 0$; *and* σ $=$ variance of $(Y|X)$.

According to the least squares principle the estimates of regression parameters are those which minimize the residual sum of squares $e^T e$. Hence

$$e^T e = (Y - Xb)^T (Y - Xb)$$

is differentiated with respect to b, and the resulting expression is set equal to zero. This gives:

$$X^T Xb = X^T Y$$

which are called the normal equations. Multiplying both sides with $(X^T X)^{-1}$ leads to an explicit expression for b:

$$\frac{b}{(p*1)} = \frac{(X^T X)^{-1}}{(p*n)(n*p)} \frac{X^T}{(p*n)} \frac{Y}{(n*1)}$$

Note that the independent variables should be chosen such that none of these is a linear combination of other independent variables. The properties of the estimator b:

$$Cov(b) = \sigma_\varepsilon^2 (X^T X)^{-1}$$

The total adjusted sum of squares $Y^T Y$ can be partitioned into an explained part due to regression and an unexplained part about regression, as follows:

$$Y^T Y = b^T X^T Y + e^T e.$$

where $(Xb)^T Y$ = sum of squares due to regression; $e^T e$ = sum of squares about regression. This equation states:

Total sum of squares about mean = regression sum of squares + residual sum of squares

The mean squares values of the right hand side terms in equation are obtained by dividing the sum of squares by their corresponding degrees of freedom. If b is a (p × 1)-column vector, i.e. there are p-independent variables in regression, then the regression sum of squares has p- degrees of freedom. Since the total sum of squares has (n-1)-degrees of freedom (note: 1 degree of freedom is lost due to the estimation of ȳ), it follows by subtraction that the residual sum of squares has (n-1-p)-degrees of freedom. It can be shown that the residual mean square S_e^2

$$S_e^2 = \frac{e^T e}{n-1-p}$$

Is an unbiased estimate of σ_ε^2. The estimate se of σ_ε is the standard error of estimate.

The analysis of variance (ANOVA) table summarizes the sum of squares quantities.

Source	Sum of squares	Degrees of freedom
Total	$S_Y = Y^T Y$	n
Mean	$n\bar{Y}^2$	1
Regression	$b^T X^T Y - n\bar{Y}^2$	p-1
Residual	$Y^T Y - b^T X^T Y$	n-p

As for the simple linear regression a measure for the quality of the regression equation is the coefficient of determination, defined as the ratio of the explained or regression sum of squares and the total adjusted sum of squares.

$$R_m^2 = \frac{b^T X^T Y}{Y^T Y} = 1 - \frac{e^T e}{Y^T Y}$$

Confidence Intervals on the Regression Line

To place confidence limits on Y_0 where $Y_0 = X_0 b$ it is necessary to have an estimate for the variance of $\hat{Y}_0$. Considering Cov(b) as given the variance Var($\hat{Y}_0$) is given by:

$$Var\left(\hat{Y}_0\right) = S_e^2 X_0 \left(X^T X\right)^{-1} X_0^T$$

The confidence limits for the mean regression equation are given by

$$CL_\pm = X_0 b \pm t_{1-a/2,n-p}\sqrt{Var\left(\hat{Y}_0\right)}$$

Coefficient of Determination (R^2)

Let $Z_{i,j} = \left(X_{i,j} - \overline{x}_j \right) / S_j$

where $\overline{x}_j$ and S_j are the mean and standard deviation of the jth independent variable. The correlation matrix is:

$$R = Z^T Z / (n-1) = \left[R_{i,j} \right]$$

where $R_{i,j}$ is the correlation between the ith and j^{th} independent variables. R is a symmetric matrix since $R_{i,j} = R_{j,i}$. The coefficient of determination is defined as

R^2 = Sum of squares due to regression / Sum of squares about mean

Or
$$R^2 = \left(b^T X^T Y - n\overline{Y}^2 \right) / \left(Y^T Y - n\overline{Y}^2 \right)$$

$$\left(L_{\hat{b}_i}, U_{\hat{b}_i} \right) = \left(\hat{b}_i - t_{(1-\alpha/2)(n-p)} S_{\hat{b}_i}, \hat{b}_i + t_{(1-\alpha/2)(n-p)} S_{\hat{b}_i} \right)$$

Here b^T is the transpose of vector b of size $(1xp)$, and Y^T is the transpose of vector Y of size $(1xn)$. Let the residual error be $\varepsilon = Y - Xb$. R^2 is the part of the total sum of squares conceted for mean that is explain by the regression equation. It ranges between 0 and 1 and closer it is to 1, the better is the regression.

Inferences on Regression Coefficients

(i) Confidence intervals on b_i

Assuming that the model is correct, the quantity $\hat{b}_i / S_{\hat{b}_i}$ follows a t-distribution with $(n-p)$ degrees of freedom. The confidence intervals on b_i are given as

(ii) Test of hypothesis concerning b_i

The hypothesis that the ith variable is not contributing significantly to explaining the variation in the dependent variable is equivalent to testing the hypothesis $H_0 : b_i = b_o$ versus $H_a : b_i \neq b_o$ The test is conducted by computing:

$$t = \left(\hat{b}_i - b_o \right) / S_{\hat{b}_i}$$

Null hypothesis H_0 is rejected if $|t| > t_{(1-\alpha/2),(n-p)}$. If this hypothesis is accepted, it is advisable to delete the concerned variable from the regression model.

Significance of the Overall Regression

The null hypothesis $H_0 : b_1 = b_2 = ... b_p = 0$ versus   at least one of these b's is not zero is used to test whether the regression equation is able to explain a significant amount of variation of Y or not. The ratio of the mean square error due to regression to the residual mean square has an F distribution with $p - l$ and $n - p$ degrees of freedom. Hence, the hypothesis is tested by computing the test statistic:

$$F = \frac{\left(b^T X^T Y - n\bar{Y}^2\right)(p-1)}{\left(Y'Y - \hat{b}'X'Y\right)/(n-p)}$$

H_0 is rejected if F exceeds the critical value $F_{(1-\alpha),(p-1),(n-p)}$.

Confidence Intervals on Regression Line:

To put the confidence limits on $Y_k = X_k b$, it is necessary to estimate the variance of $\hat{y}_k$. This is given by

$$S_{\hat{Y}_k}^2 = S^2 X_k \left(X'X\right)^{-1} X'_k$$

where $\left(L,U\right) = \left\{ \hat{Y}_k - t_{(1-\alpha/2)(n-p)} S_{\hat{Y}_k}, \hat{Y}_k + t_{(1-\alpha/2)(n-p)} S_{\hat{Y}_k} \right\}$

Confidence Intervals on Individual Predicted Value of Y

$$\hat{Y}_K = X_K \hat{b}$$

$$\left(L',U'\right) = \left\{ \hat{Y}_k - t_{(1-\alpha/2)(n-p)} S'_{\hat{Y}_k}, \hat{Y}_k + t_{(1-\alpha/2)(n-p)} S'_{\hat{Y}_k} \right\}$$

$$S'^2_{\hat{Y}_k} = S^2 \left[1 + X_k \left(X'X\right)^{-1} X'_k \right]$$

Example: Table contains rainfall for the months of July and August and discharge for the August month for a catchment. Estimate the parameters of linear regression and multiple linear regression and find out if there is an advantage in using multiple linear regression in this case.

Table: Data and computations for multiple linear regression example

YEAR	RF-JUL (MCM)	RF-AUG (MCM)	Obs Q Aug (MCM)	Comp. Q by Lin Reg (Q_L)	$(Qob-Q_L)^2$	Comp. Q by Mult. Lin Reg (Q_M)	$(Qob-Q_M)^2$
1982	5020.04	15664.05	5996.939	6830.0	694015.6	6873.0	767532
1983	7980.13	6546.24	2557.916	3263.6	497987.7	3572.3	1028983
1984	3002.36	13086.63	4395.515	5821.9	2034467.0	4736.5	116242
1985	8572.75	7532.13	5725.02	3649.2	4308915.7	4314.0	1990914
1986	5242.03	5799.34	2532.373	2971.4	192787.2	2045.2	237329
1987	6311.05	9522.80	2774.517	4427.9	2733589.2	4353.5	2493329
1988	6040.00	7285.46	4163.013	3552.7	372427.7	3123.1	1081472
1989	1597.33	6922.49	2046.694	3410.8	1860702.7	1068.8	956235

1990	8561.71	6889.43	4190.084	3397.8	627652.7	3988.7	40541
1991	7153.31	12566.82	6107.452	5618.5	239036.6	6227.3	14365
1992	5623.67	10263.08	5145.44	4717.4	183188.8	4433.0	507510
1993	4233.30	7108.91	2300.774	3483.7	1399281.8	2273.2	759
1994	13076.88	10472.23	8994.085	4799.2	17596705.8	7679.9	1727088
1995	6843.64	8068.47	3695.11	3859.0	26865.5	3852.6	24788
1996	7819.49	9330.16	4870.4	4352.5	268196.6	4893.4	531
1997	9403.82	7424.92	3943.455	3607.3	113005.8	4610.9	445541
1998	7040.85	8306.55	3801.727	3952.1	22624.6	4054.5	63883
1999	7380.56	9987.30	5895.899	4609.6	1654653.4	5036.2	739043
2000	8620.28	4283.79	1501.445	2378.6	769480.6	2713.5	1469074
2001	9113.46	5071.52	2670.739	2686.8	256.8	3314.4	414338
2002	1296.93	11168.68	3192.95	5071.7	3529547.4	3060.5	17531
2003	7493.84	7784.62	3708.33	3748.0	1572.8	3985.1	76593
Sum	147427.41	191085.61	90209.88	90209.88	3.91E+07	9.02E+04	1.42E+07
Average	6701.25	8685.71	4100.45	4100.449			

Solution: Using the data given in the table, linear regression equation of the following form was established between the rainfall and observed discharge for the August month.

$Q_A = a + bR_A$ where Q_A = discharge for August and R_A = rainfall for August.

The parameters a and b were estimated to yield the following

$$\text{equation: } Q_A = 703.05 + 0.391 R_A$$

Coefficient of determination $R^2 = 1 - 3.91 * 10^7 / 6.33 * 10^7 = 0.382$

Next, discharge for the August month was computed by using the above equation and the sum of square of residuals turned out to be $3.91 * 10^7$.

In case of multiple linear regression, the independent variables were the rainfall for the month July and August and dependent variable as the discharge for August. Regression equation of the following form was envisaged

$$Q_A = a + b_1 R_J + b_2 R_A$$

where R_J = rainfall for July month.

After computations, the following regression equation was obtained.

$$Q_A = -3058.24 + 0.42R_J + 0.50R_A$$

Coefficient of determination $R^2 = 1 - 1.42*10^7/6.33*10^7 = 0.78$.

The discharge for August was computed by LR and MLR equations and the sum of squares of errors were computed. The values were $3.91*10^7$ for LR and $1.42*10^7$ for MLR. When these values are compared along with R^2 for the two cases, it can be concluded that MLR gives much improved estimates of the discharge compared to LR.

Comments on Multiple Regression

As demonstrated through the example, multiple regression is suitable in situations where one dependent variable and several independent variables are available and it is desired to fit a linear model containing all of the significant independent variables. Two questions that may be asked are: (1) Is a linear model suitable for the problem? and (2) which independent variables should be included in the model?

The first question can be answered by plotting the data and computing the statistical performance indices. Regarding the second question, a factor to be considered in selection of the variables is that in most cases, the independent variables are not statistically independent and are correlated. Hence while using regression analysis, the correlation matrix should be computed in the beginning.

If the regression equation contains independent variables that are highly correlated among themselves then (besides the mathematical difficulties in determining the coefficients) the interpretation of the regression coefficients becomes difficult. Many times the sign and magnitudes of coefficient of a variable may be different than what is expected if the corresponding variable is highly correlated with another independent variable in the equation.

A common practice in multiple regression analysis is to perform several regressions with the given set of data using different combinations of the independent variables. Finally the regression equation that best fits the data is selected. A commonly used criterion for the best fit is to select the equation that gives the smallest value of the sum of squares of errors. A scatter plot of observed and computed values is always helpful.

All of the variables included in a regression equation should make a significant contribution to the model unless there is an overriding reason (theoretical or intuitive) for retaining a particular variable. The variables retained should have physical significance. If two variables are equally significant when used alone and are highly correlated than the one that is easiest to obtain should be used.

The number of variables retained in regression should be small compared to the number of observation and should not exceed 25 to 30 percent of the number of observations. This is a rule of thumb to avoid "over-fitting" whereby oscillations in the prediction may occur.

Stepwise Regression

Stepwise regression is a procedure that is commonly used to select the "best" regression equation (by including only relevant variables) from amongst a number of independent variables. This approach consists of building the regression equation by adding one variable at a time. At each step, all the variables in the regression equation are examined for significance and at any stage, a particular variable is removed if it is no longer explaining a significant variation of the dependent variable.

To begin with, the first variable to be added in the regression equation is the one which has the highest correlation with the dependent variable. The second variable to be added is the one that explains the largest remaining unexplained variation in the dependent variable. At this stage the first variable is tested for significance and retained or discarded depending on the results of this test. The third variable added is the one that explains the largest portion of the variation that is not explained by the two regression variables already in the equation. The variables in the equation are then tested for significance. This procedure is repeated till a situation is reached when all the variables that are not in the equation are insignificant and all the variables that are in the equation are significant. This is a very good approach to use but care must be exercised to ensure that the resulting equation is rational.

The steps of stepwise regression are:

  i.   The variable which has the highest correlation with the dependent variable is picked up as the first independent variable.

  ii. The variable which explains the largest of the residual variation in the dependent variable after the first step is added as the next variable.

  iii. Test the significance of the new variable and retain or discard it depending on the results of this test.

  iv. Repeat steps (ii) & (iii) until each of the variables that are not in the equation are found to be insignificant and all the variables in the equation are significant.

The real test of how good is the regression model (or any other model), is the ability of the model to predict the dependent variable by using the observations of the independent variables that were not used in estimating the regression coefficients. For this purpose, the data are divided into two parts. One part of the data is used to develop the model and the other part to test the model. Transformation of independent variables may significantly improve the regression relationship.

The difference between multiple and stepwise regression is that in multiple linear regression all independent variables are included in the regression model, whereas in stepwise regression, the equation is built up step by step by taking those independent variables into consideration first, which reduce the error variance most. The entry of new independent variables is continued until the reduction in the error variance falls below a certain limit. In some stepwise regression tools a distinction is made between free and forced independent variables. A forced variable will always be included into the equation no matter what error variance reduction it produces, whereas a free variable enters only if the error variance reduction criterion is met.

Transforming Non Linear Relations

Relationship between some variables may be non-linear but can be transformed to linear form so that the technique of linear regression can be applied. For example, consider that two variables X and Y are non-linearly related as follows:

$$Y = \alpha X^{\beta}$$

This non-linear relation can be linearized by logarithmic transformation of the equation

$$\text{Ln } Y = \text{Ln } \alpha + \beta \text{Ln } X$$

$$\text{Or } A = a + b * B$$

where $A = \text{Ln } Y$, $a = \text{Ln } \alpha$, $b = \beta$, and $B = \text{Ln } X$. Now, one can use the regression technique to estimate parameters a and b and thereby α and β. In this procedure, two important points are worth noting.

a) The values of a and b are estimated by minimizing $\sum \left(A - A_{reg}\right)^2$ and not by minimizing $\sum \left(Y - Y_{reg}\right)^2$. Here A_{reg} and Y_{reg} are the value of A and Y estimated by the regression equation.

b) In the log-transformed equation, the error term is additive $\left(A = a + bB + c\right)$ which means that it is multiplicative in the original equation

$$Y = \alpha X^{\beta} \varepsilon$$

The errors are related as $c = \ln \varepsilon$. Hence, the assumptions in hypothesis testing and confidence intervals should be valid for c.

In some cases, it has been observed that the log transformed data follows the regression assumptions more closely than the original data. The standard regression is based on a constant absolute error along the regression line whereas the normal equations for a logarithmic transformation are based on a constant percentage error along the regression line.

Dependent and Independent Variables

In mathematical modelling and statistical modelling, there are dependent and independent variables. The models investigate how the former depend on the latter. The dependent variables represent the output or outcome whose variation is being studied. The independent variables represent inputs or causes, i.e. potential reasons for variation. Models test or explain the effects that the independent variables have on the dependent variables. Sometimes, independent variables may be included for other reasons, such as for their potential confounding effect, without a wish to test their effect directly.

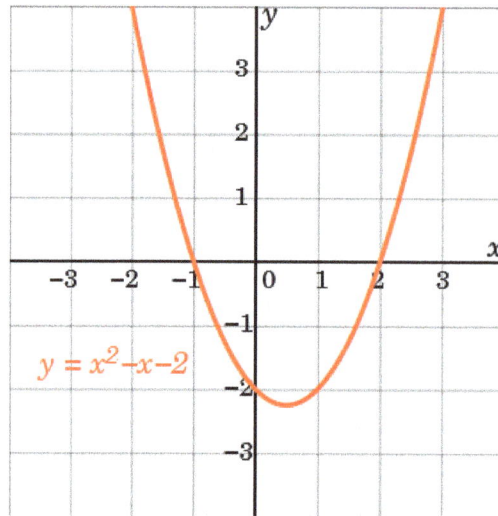

In calculus, a function is typically graphed with the horizontal axis representing the independent variable and the vertical axis representing the dependent variable. In this function, y is the dependent variable and x is the independent variable.

Mathematics

In mathematics, a function is a rule for taking an input (usually number or set of numbers) and providing an output (which is also usually a number). A symbol that stands for an arbitrary input is called an independent variable, while a symbol that stands for an arbitrary output is called a dependent variable. The most common symbol for the input is x, and the most common symbol for the output is y; the function itself is commonly written $y = f(x)$.

It is possible to have multiple independent variables and/or multiple dependent variables. For instance, in multivariable calculus, one often encounters functions of the form $z = f(x, y)$, where z is a dependent variable and x and y are independent variables. Functions with multiple outputs are often written as vector-valued functions.

In Set Theory, a function between a set X and a set Y is a subset of the Cartesian product such that every element of X appears in an ordered pair with exactly one element of Y. In this situation, a symbol representing an element of X may be called an independent variable and a symbol representing an element of Y may be called a dependent variable, such as when X is a manifold and the symbol x represents an arbitrary point in the manifold. However, many advanced textbooks do not distinguish between dependent and independent variables.

Statistics

In an experiment, the dependent variable is the event expected to change when the independent variable is manipulated.

In data mining tools (for multivariate statistics and machine learning), the depending variable is assigned a *role* as target variable (or in some tools as *label attribute*), while a dependent variable may be assigned a role as *regular variable*. Known values for the target variable are provided for the training data set and test data set, but should be predicted for other data. The target variable is used in supervised learning algorithms but not in non-supervised learning.

Modelling

In mathematical modelling, the dependent variable is studied to see if and how much it varies as the independent variables vary. In the simple stochastic linear model $y_i = a + bx_i + e_i$ the term y_i is the ith value of the dependent variable and x_i is i^{th} value of the independent variable. The term e_i is known as the "error" and contains the variability of the dependent variable not explained by the independent variable.

With multiple independent variables, the expression is: $y_i = a + bx_1 + bx_2 + ... + bx_n + e_i$, where *n* is the number of independent variables.

Simulation

In simulation, the dependent variable is changed in response to changes in the independent variables.

Statistics Synonyms

Depending on the context, an independent variable is sometimes called a "predictor variable", "regressor", "controlled variable", "manipulated variable", "explanatory variable", "exposure variable", "risk factor", "feature" (in machine learning and pattern recognition) or "input variable."

Depending on the context, a dependent variable is sometimes called a "response variable", "regressand", "predicted variable", "measured variable", "explained variable", "experimental variable", "responding variable", "outcome variable", "output variable" or "label".

"Explanatory variable" is preferred by some authors over "independent variable" when the quantities treated as independent variables may not be statistically independent or independently manipulable by the researcher. If the independent variable is referred to as an "explanatory variable" then the term "response variable" is preferred by some authors for the dependent variable.

"Explained variable" is preferred by some authors over "dependent variable" when the quantities treated as "dependent variables" may not be statistically dependent. If the dependent variable is referred to as an "explained variable" then the term "predictor variable" is preferred by some authors for the independent variable.

Variables may also be referred to by their form: continuous, binary/dichotomous, nominal categorical, and ordinal categorical, among others.

Other Variables

A variable may be thought to alter the dependent or independent variables, but may not actually be the focus of the experiment. So that variable will be kept constant or monitored to try to minimise its effect on the experiment. Such variables may be designated as either a "controlled variable", "control variable", or "extraneous variable".

Extraneous variables, if included in a regression as independent variables, may aid a researcher with accurate response parameter estimation, prediction, and goodness of fit, but are not of substantive interest to the hypothesis under examination. For example, in a study examining the effect of post-secondary education on lifetime earnings, some extraneous variables might be gender, ethnicity, social class, genetics, intelligence, age, and so forth. A variable is extraneous only when it can be assumed (or shown) to influence the dependent variable. If included in a regression, it can improve the fit of the model. If it is excluded from the regression and if it has a non-zero covariance with one or more of the independent variables of interest, its omission will bias the regression's result for the effect of that independent variable of interest. This effect is called confounding or omitted variable bias; in these situations, design changes and/or statistical control is necessary.

Extraneous variables are often classified into three types:

1. Subject variables, which are the characteristics of the individuals being studied that might affect their actions. These variables include age, gender, health status, mood, background, etc.

2. Blocking variables or experimental variables are characteristics of the persons conducting the experiment which might influence how a person behaves. Gender, the presence of racial discrimination, language, or other factors may qualify as such variables.

3. Situational variables are features of the environment in which the study or research was conducted, which have a bearing on the outcome of the experiment in a negative way. Included are the air temperature, level of activity, lighting, and the time of day.

In modelling, variability that is not covered by the independent variable is designated by and is known as the "residual", "side effect", "error", "unexplained share", "residual variable", or "tolerance".

References

- Armstrong, J. Scott (2012). "Illusions in Regression Analysis". International Journal of Forecasting (forthcoming). 28 (3): 689. doi:10.1016/j.ijforecast.2012.02.001

- Mogull, Robert G. (2004). Second-Semester Applied Statistics. Kendall/Hunt Publishing Company. p. 59. ISBN 0-7575-1181-3

- Waegeman, Willem; De Baets, Bernard; Boullart, Luc (2008). "ROC analysis in ordinal regression learning". Pattern Recognition Letters. 29: 1–9. doi:10.1016/j.patrec.2007.07.019

- Yule, G. Udny (1897). "On the Theory of Correlation". Journal of the Royal Statistical Society. Blackwell Publishing. 60 (4): 812–54. doi:10.2307/2979746. JSTOR 2979746

- Ronald A. Fisher (1954). Statistical Methods for Research Workers (Twelfth ed.). Edinburgh: Oliver and Boyd. ISBN 0-05-002170-2

- Galton, Francis (1989). "Kinship and Correlation (reprinted 1989)". Statistical Science. Institute of Mathematical Statistics. 4 (2): 80–86. doi:10.1214/ss/1177012581. JSTOR 2245330

- Tofallis, C. (2009). "Least Squares Percentage Regression". Journal of Modern Applied Statistical Methods. 7: 526–534. doi:10.2139/ssrn.1406472. SSRN 1406472

- Fotheringham, A. Stewart; Brunsdon, Chris; Charlton, Martin (2002). Geographically weighted regression: the analysis of spatially varying relationships (Reprint ed.). Chichester, England: John Wiley. ISBN 978-0-471-49616-8

- Pearson, Karl; Yule, G.U.; Blanchard, Norman; Lee,Alice (1903). "The Law of Ancestral Heredity". Biometrika. Biometrika Trust. 2 (2): 211–236. doi:10.1093/biomet/2.2.211. JSTOR 2331683

- Francis Galton. "Regression Towards Mediocrity in Hereditary Stature," Journal of the Anthropological Institute, 15:246-263 (1886). (Facsimile at: [1])

- Chiang, C.L, (2003) Statistical methods of analysis, World Scientific. ISBN 981-238-310-7 - page 274 section 9.7.4 "interpolation vs extrapolation"

- Aldrich, John (2005). "Fisher and Regression". Statistical Science. 20 (4): 401–417. doi:10.1214/088342305000000331. JSTOR 20061201

- Good, P. I.; Hardin, J. W. (2009). Common Errors in Statistics (And How to Avoid Them) (3rd ed.). Hoboken, New Jersey: Wiley. p. 211. ISBN 978-0-470-45798-6

- Fotheringham, AS; Wong, DWS (1 January 1991). "The modifiable areal unit problem in multivariate statistical analysis". Environment and Planning A. 23 (7): 1025–1044. doi:10.1068/a231025

Hydrologic Simulation Models in Hydrology

A hydrologic model represents a hydrologic cycle. Hydrologic models can be categorized into two models, stochastic models and process-based models. Hydrological Simulation Program-Fortran (HSPF), MIKE, etc. are some of the hydrologic models explored in this chapter. The aspects elucidated in this chapter are of vital importance, and provide a better understanding of hydrologic models.

Hydrologic Models

Watershed Classification

Watershed (ha)	Classification
50,000-2,00,000	Watershed
10,000-50,000	Sub-watershed
1,000-10,000	Milli-watershed
100-1,000	Micro-watershed
10-100	Mini-watershed

Hydrologic Simulation Model

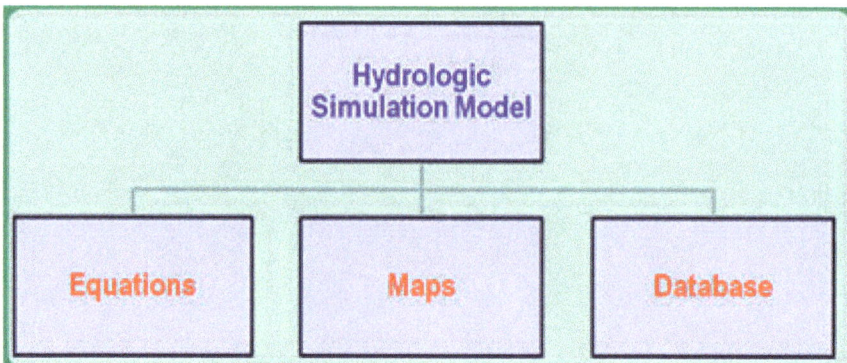

A hydrologic simulation model is composed of three basic elements, which are:

(1) Equations that govern the hydrologic processes,

(2) Maps that define the study area and

(3) Database tables that numerically describe the study area and model parameters.

A hydrological simulation model can also be defined here as a mathematical model aimed at synthesizing a (continuous) record of some hydrological variable Y, for a period T, from available concurrent records of other variables X, Z... In contrast, a hydrological forecasting model is aimed at synthesizing a record of a variable Y (or estimating some of its states) in an interval ΔT, from available records of the same variable Y and/or other variables X, Z, ..., in an immediately preceding period T. A hydrological simulation model can operate in a "forecasting mode" if estimates of the records of the independent variables (predictors) X, Z, ..., for the forecast interval ΔT are available through an independent forecast. Then the simulation model, by simulating a record of the dependent variable, $[Y]_{\Delta T}$ will in fact produce its forecast.

In short, a hydrological simulation model works in a forecasting mode whenever it uses forecasted rather than observed records of the independent variables.

A Typical Watershed Delineation Model

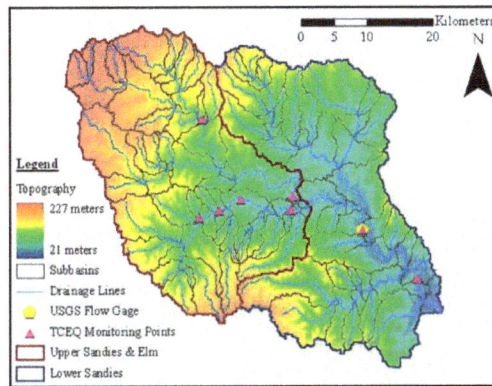

Classification of Hydrologic Models

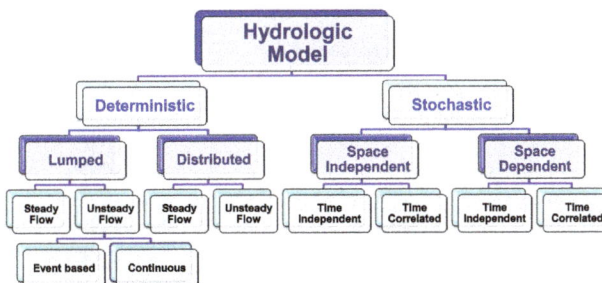

Spatial Scaling of Models

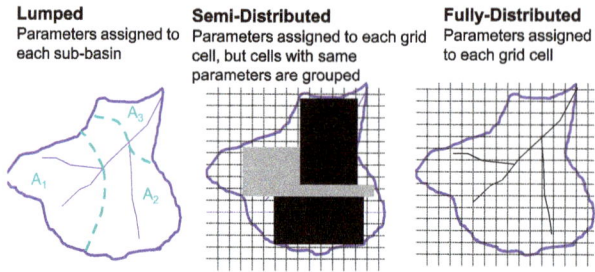

Lumped
Parameters assigned to each sub-basin

Semi-Distributed
Parameters assigned to each grid cell, but cells with same parameters are grouped

Fully-Distributed
Parameters assigned to each grid cell

Parameters of Watershed

1. Size

2. Shape

3. Physiography

4. Climate

5. Drainage

6. Land Use

7. Vegetation

8. Geology and Soils

9. Hydrology

10. Hydrogeology

11. Socioeconomics

Flowchart of Simple Watershed Model (McCuen, 1989)

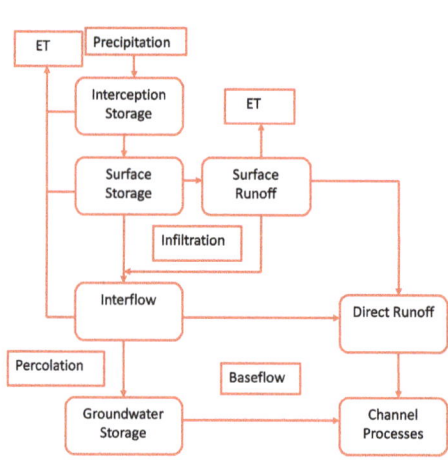

Strengths of Watershed Models

Diversity of the Current Generation of Models:

There exists a multitude of watershed models, and their diversity is so large that one can easily find more than one watershed model for addressing any practical problem.

Comprehensive Nature

Many of the models can be applied to a range of problems.

Reasonable modeling of physical phenomena

In many cases models mimic reasonably well the physics of the underlying hydrologic processes in space and time.

Distributed in Space and Time

Many models are distributed in space and time.

Multi-disciplinary nature

Several of the models attempt to integrate with hydrology :

a) Ecosystems and ecology,

b) Environmental components,

c) Biosystems,

d) Geochemistry,

e) Atmospheric sciences and

f) Coastal processes

This reflects the increasing role of watershed models in tackling environmental and ecosystems problems.

Deficiencies of Watershed Models

The most ubiquitous deficiencies of the models are:

- Lack of user-friendliness,

- Large data requirements,

- Lack of quantitative measures of their reliability,

- Lack of clear statement of their limitations, and

- Lack of clear guidance as to the conditions for their applicability.

Also, some of the models cannot be embedded with social, political, and environmental systems.

> Although watershed models have become increasingly more sophisticated, there is a long way to go before they become "household" tools.

Hydrologic Models

Model Type	Example of Model
Lumped parameter	Snyder or Clark UH
Distributed	Kinematic wave
Event	HEC-1,HEC-HMS,SWMM,SCS,TR-20
Continuous	Stanford Model,SWMM,HSPF,Storm
Physically based	HEC-1,HEC-HMS,SWMM,HSPF
Stochastic	Synthetic streamflows
Numerical	Kinematic or dynamic wave models
Analytical	Rational Method,Nash IUH
Models	Application Areas
HEC-HMS	Design of drainage systems,quantifying the effect of land use change on flooding
National Weather Service (NWS)	Flood forecasting
Modular Modeling System (MMS)	Water resources planning and management works
University of British Columbia (UBC) & WATFLOOD	Hydrologic simulation
Runoff-Routing model (RORB) &WBN	Flood forecasting, drainagedesign,and evaluating the effect of land use change
TOPMODEL & SHE	Hydrologic analysis
HBV	Flow forecasting

HSPF (Hydrological Simulation Program-Fortran)

Commercial successor of the Stanford Watershed Model (SWM-IV) (Johanson et al., 1984):

- Water-quality considerations
- Kinematic Wave routing
- Variable Time Steps

> **HSPF** is a deterministic, lumped parameter, physically based, continuous model for simulating the water quality and quantity processes that occur in watersheds and in a river network.

HSPF incorporates watershed-scale ARM (Agricultural Run-off Management) and NPS (Non-Point Source) models into a basin-scale analysis framework based on fate and transport of pollutants in 1-D stream channels.

Data Requirements of HSPF

- Rainfall
- Infiltration
- Baseflow
- Streamflow
- Soils
- Landuse

HSPF is one of the most complex hydrologic models which simulates:

- Infiltration: Philip's equation, a physically based method which uses an hourly time step.
- Streamflow: Chezy – Manning's equation.

HSPF can simulate temporal scales ranging from minutes to days. Due to its flexible modular design, HSPF can model systems of varying size and complexity.

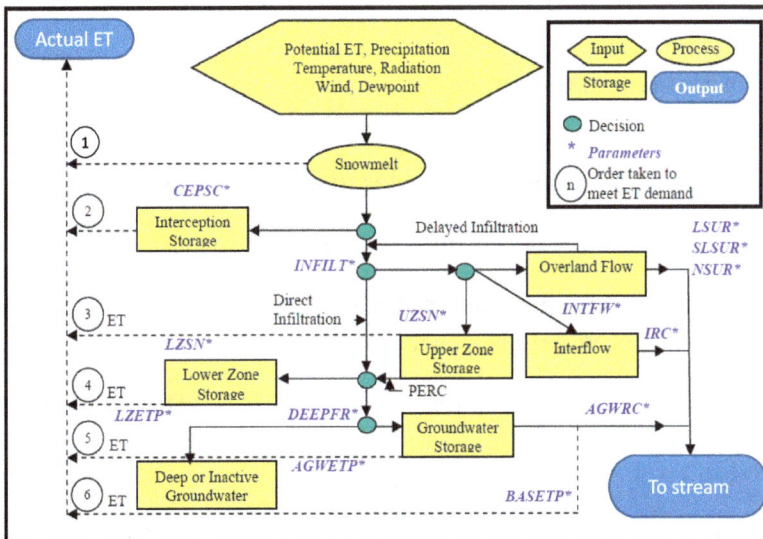

Stanford Watershed Model(AquaTerra, 2005)

CEPSC   :   interception storage capacity

LSUR   :   length of the overland flow plane

SLSUR   :   slope of the overland flow plane

NSUR   :   Manning's roughness of the land surface

INTFW   :   interflow inflow

INFLIT   :   index to the infiltration capacity of the soil

UZSN   :   nominal capacity of the upper-zone storage

IRC   :   interflow recession constant

LZSN   :   nominal capacity of the lower-zone storage

LZETP   :   lower-zone evapotranspiration

AGWRC   :   basic ground-water recession rate

AGWETP :   fraction of remaining potential evapotranspiration that can be satisfied
                from active ground-water storage

KVARY   :   indication of the behavior of ground-water recession flow

DEEPFR   :   fraction of ground-water inflow that flows to inactive ground water

BASETP   :   fraction of the remaining potential evapotranspiration that can be satis-
                fied from base flow

HEC Models

Modeling of the rainfall-runoff process in a watershed based on watershed physiographic data

- a variety of modeling options in order to compute UH for basin areas.

- a variety of options for flood routing along streams.

- capable of estimating parameters for calibration of each basin based on comparison of computed data to observed data.

1. HEC-GridUtil 2.0   2.   HEC-GeoRAS 10 (EAP)

3. HEC-GeoHMS 10 (EAP)   4.   HEC-GeoEFM 1.0

5. HEC-SSP 2.0   6.   SnoTel 1.2 Plugin

7. HEC-HMS 3.5   8.   HEC-FDA 1.2.5a

9. HEC-DSSVue 2.0.1   10. HEC-RAS 4.1

11. HEC-DSS Excel Add-In   12. HEC-GeoDozer 1.0

13. HEC-EFM 2.0   14. HEC-EFM Plotter 1.0

15. HEC-ResSim 3.0   16. HEC-RPT 1.1

17. HEC-GridUtil 2.0

HEC-GridUtil 2.0

HEC-GridUtil is designed to provide viewing, processing, and analysis capabilities for gridded data sets stored in HEC-DSS format (Hydrologic Engineering Center's Data Storage System).

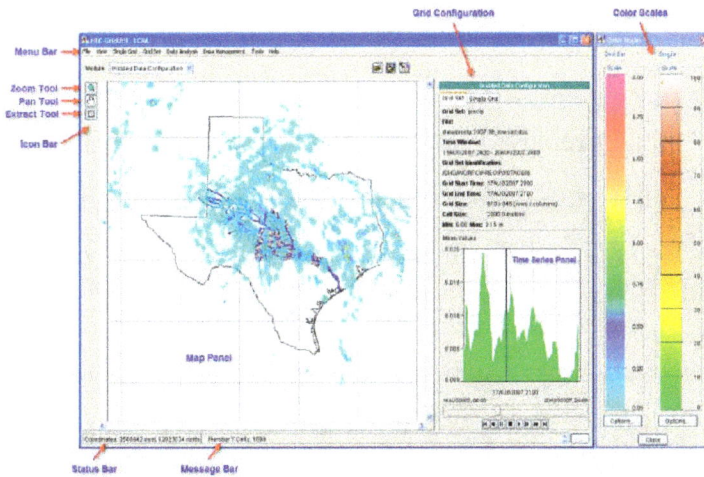

HEC-GeoRAS 10 (EAP)

GIS extension → a set of procedures, tools, and utilities for the preparation of GIS data for import into HEC-RAS and generation of GIS data from RAS output.

ArcGIS w/ extensions:

- 3D & Spatial Analyst
- HEC-GeoHMS
- HEC-GeoRAS

HEC-RAS

It is a computr program that simulates water surface profile of a stream reach.

Data Requirements

- Triangular Irregular Network (TIN)

- DEM (high resolution)
  - use stds2dem.exe if downloading from USGS

- Land Use / Land Cover
  - Manning's Coefficient

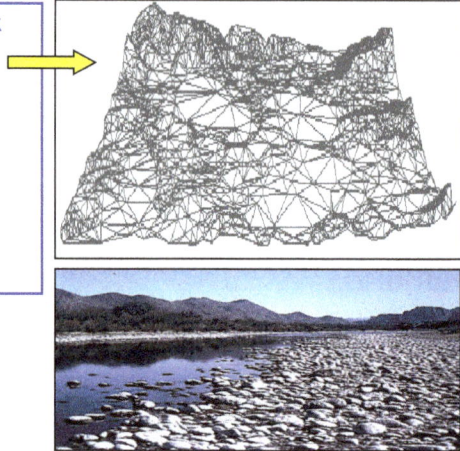

Major Functions of GeoRAS

It is an interface between ArcView and HEC-RAS.

Functions:

- PreRAS Menu - prepares Geometry Data necessary for HEC-RAS modeling

- GeoRAS_Util Menu – creates a table of Manning's n value from land use shapefile

- PostRAS Menu – reads RAS import file; delineates flood plain; creates Velocity and Depth TINs

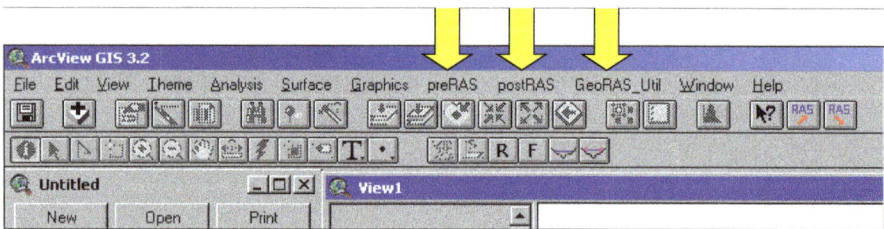

Demonstration of Capabilities

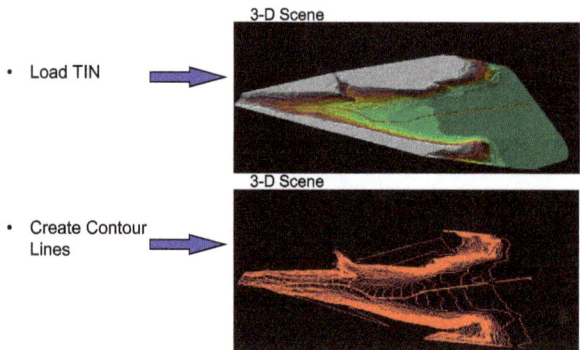

- Load TIN

3-D Scene

- Create Contour Lines

3-D Scene

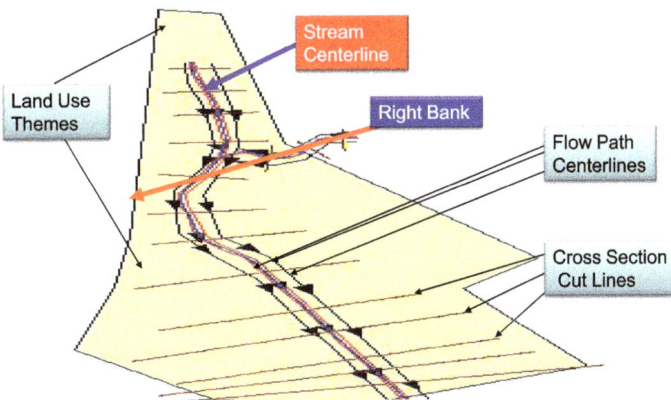

- Generate RAS GIS import file

- Open HEC-RAS and import RAS GIS file

- Complete Geometry, Hydraulic, & Flow Data

- Run Analysis

- Generate RAS Export file

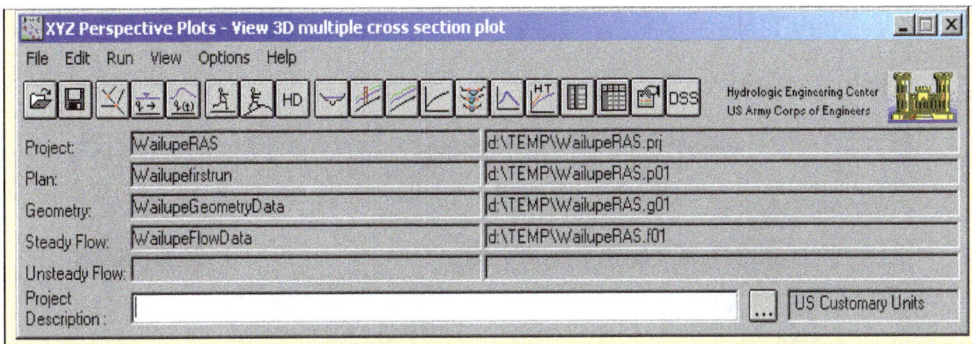

RAS GIS import file

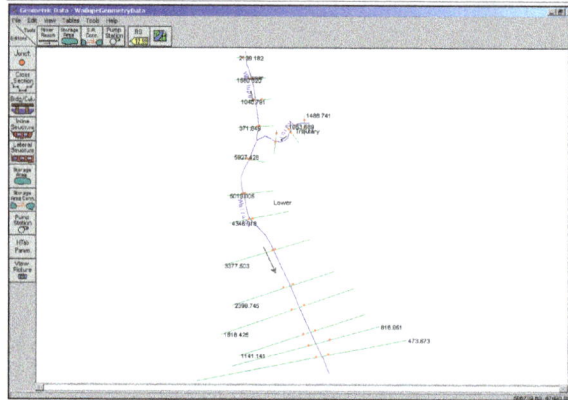

RAS GIS export file

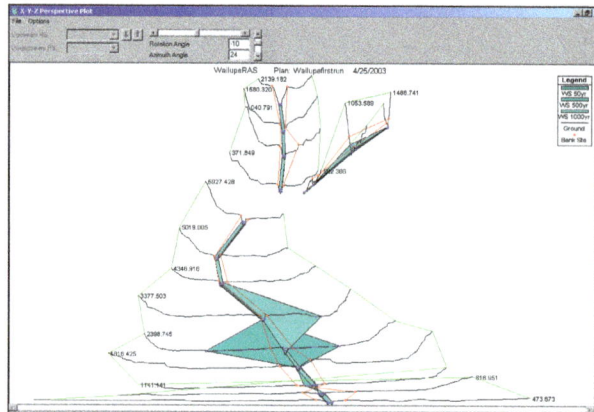

- New GIS data

- PostRAS Features

  - Water Surface TIN

  - Floodplain Delineation – polygon & grid

  - Velocity TIN

  - Velocity Grid

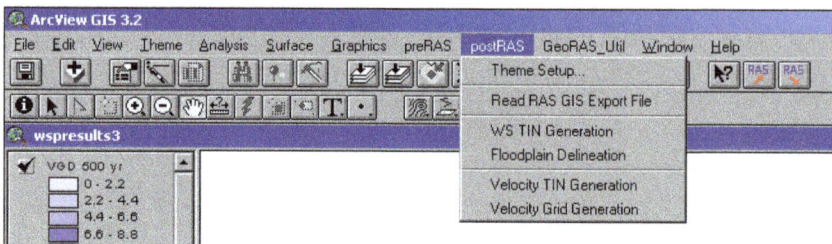

Floodplain Delineation (3-D Scene)

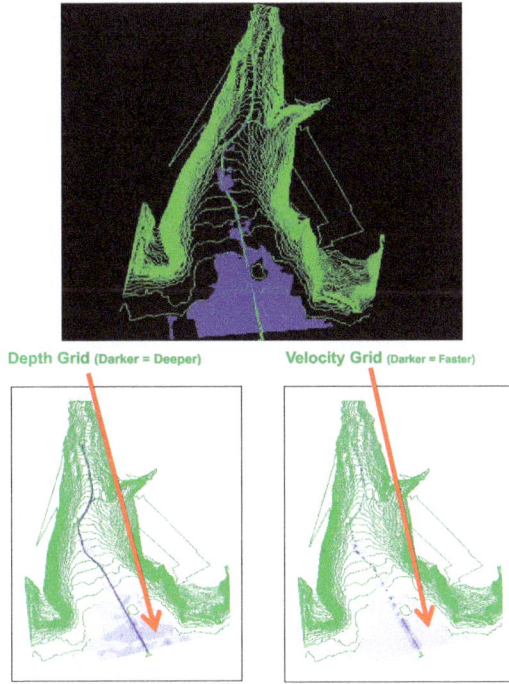

Depth Grid (Darker = Deeper)   Velocity Grid (Darker = Faster)

Employing ArcView, GeoRAS, and RAS for Main Channel Depth Analysis (1968)

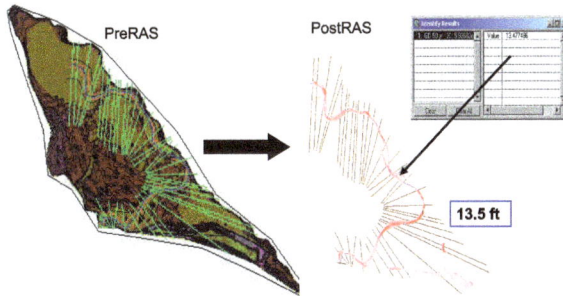

PreRAS   PostRAS   13.5 ft

Employing ArcView, GeoRAS, and RAS for Main Channel Depth Analysis (1988)

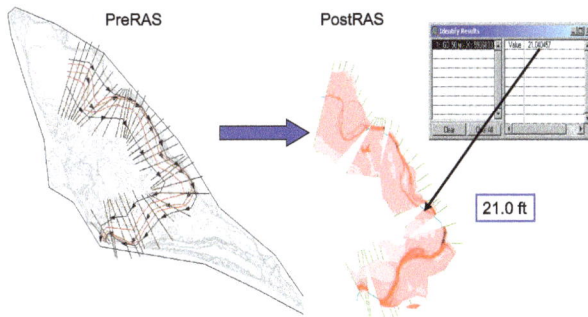

PreRAS   PostRAS   21.0 ft

HEC-HMS simulates rainfall-runoff for the watershed

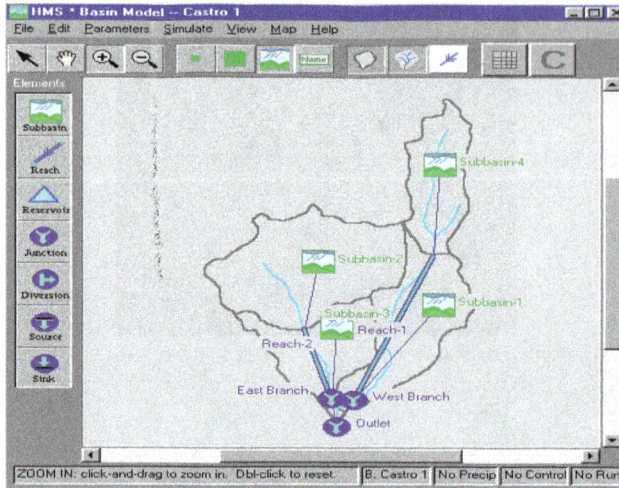

HEC-HMS Background

Purpose of HEC-HMS

- Improved User Interface, Graphics, and Reporting
- Improved Hydrologic Computations
- Integration of Related Hydrologic Capabilities

Importance of HEC-HMS

- Foundation for Future Hydrologic Software
- Replacement for HEC-1

Ease of Use

- projects divided into three components
- user can run projects with different parameters instead of creating new projects
- hydrologic data stored as DSS files
- capable of handling NEXRAD-rainfall data and gridded precipitation

HEC-HMS

Three components

- Basin model-contains the elements of the basin, their connectivity, and runoff parameters.

- Meteorologic Model - contains the rainfall and evapotranspiration data.

- Control Specifications-contains the start/stop timing and calculation intervals for the run.

Project Definition

- It may contain several basin models, meteorological models, and control specifications.

- It is possible to select a variety of combinations of the three models in order to see the effects of changing parameters on one sub-basin.

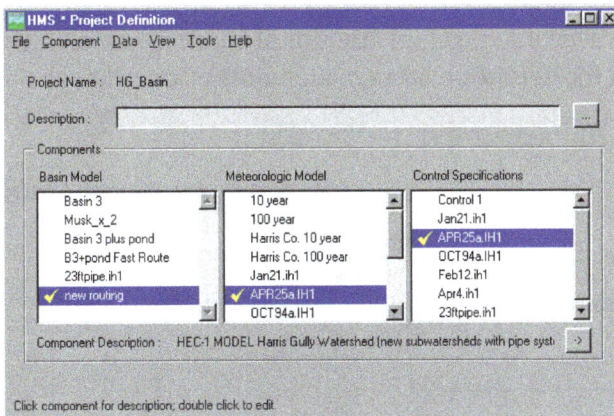

- GUI supported.

- Click on elements from left and drag into basin area.

- Works well with GIS imported files.

- Actual locations of elements do not matter, just connectivity and runoff parameters.

1. Basin Model Elements

- subbasins- contains data for subbasins (losses, UH transform, and baseflow)

- reaches- connects elements together and contains flood routing data

- junctions- connection point between elements

- reservoirs- stores runoff and releases runoff at a specified rate (storage-discharge relation)

- sinks- has an inflow but no outflow

- sources- has an outflow but no inflow

- diversions- diverts a specified amount of runoff to an element based on a rating curve - used for detention storage elements or overflows.

2. Basin Model Parameters

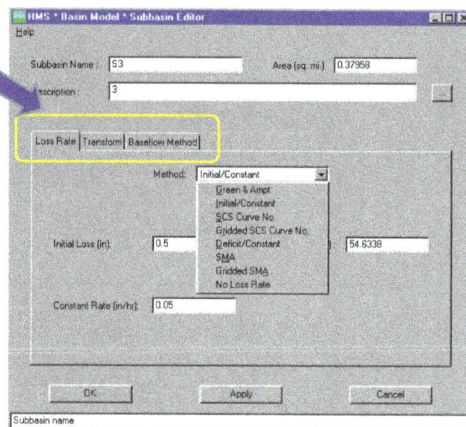

2a) Abstractions (Losses)	2b) Transformation
1. Interception Storage	1. Unit Hydrograph
2. Depression Storage	2. Distributed Runoff
3. Surface Storage	3. Grid-Based Transformation
4. Evaporation	**Methods:**
5. Infiltration	a. Clark
6. Interflow	b. Snyder
7. Groundwater and Base Flow	c. SCS
	d. Input Ordinates
	e. ModClark
	f. Kinematic Wave

2c) Baseflow Options

a. recession

b. constant monthly

c. linear reservoir

d. no base flow

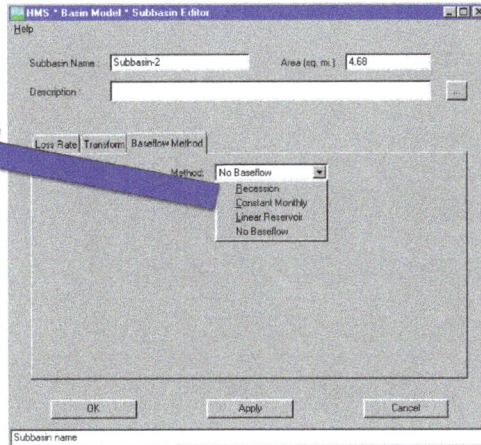

Stream Flow Routing

- Simulates movement of flood wave through stream reach.

- Accounts for storage and flow resistance.

- Allows modeling of a watershed with sub-basins.

Reach Routing

a) Simple Lag

b) Modified Puls

c) Muskingum

d) Muskingum Cunge

e) Kinematic Wave

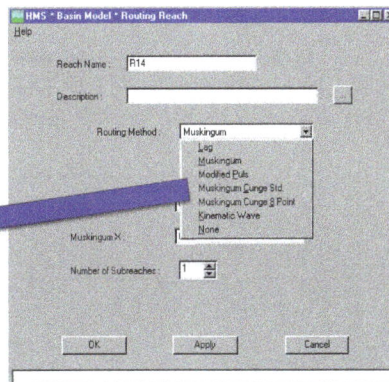

Methods for Stream Flow Routing

1. Hydraulic Methods - Uses partial form of St Venant Equations

- Kinematic Wave Method

- Muskingum-Cunge Method

2. Hydrologic Methods

- Muskingum Method

- Storage Method (Modified Puls)

- Lag Method

Reservoir Routing

- Developed Outside HEC-HMS

- Storage Specification Alternatives:

    ◊ Storage versus Discharge

    ◊ Storage versus Elevation

    ◊ Surface Area versus Elevation

- Discharge Specification Alternatives:

    ◊ Spillways, Low-Level Outlets, Pumps

    ◊ Dam Safety: Embankment Overflow, Dam Breach

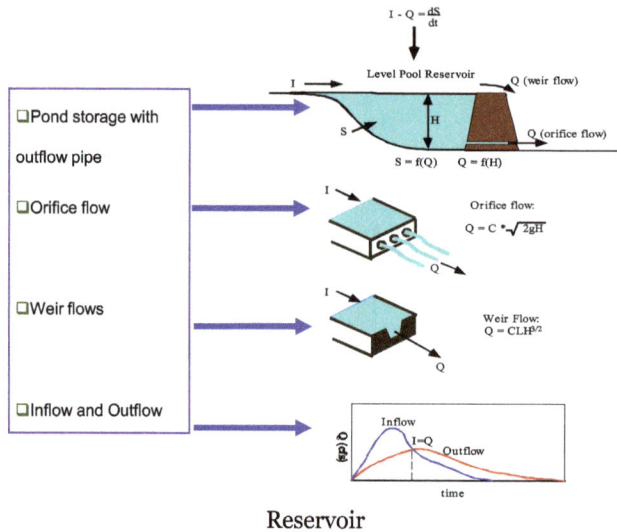

Reservoir

Reservoir Data Input

Initial conditions to be considered are:

- Inflow = Outflow

- Initial Storage Values

- Initial Outflow

- Initial Elevation

Elevation data relates to both storage/area and discharge. HEC-1 Routing routines with initial conditions and elevation data can be imported as reservoir elements.

MIKE Hydrological and Hydrodynamic Models

MIKE Zero-fication

ArcMap with MIKE ZERO

MIKE 11 GIS

- Fully integrated GIS based flood modelling
- Developed in ArcView GIS
- Pre-processing:

Floodplain schematization

- Post-processing:

> •Flood depth maps
> •Comparison maps
> •Duration maps

- Analysis with other GIS data

- Mike 11 Modules

  - HD : hydrodynamic - simulation of unsteady flow in a network of open channels. Result is time series of discharges and water levels;

  - AD : advection dispersion;

  - WQ : water quality.

Open Channel Flow

Saint Venantequations (1D)

- continuity equation (mass conservation)

- momentum equation (fluid momentum conservation)

Assumptions

- water is incompresible and homogeneous

- bottom slope is small

- flow everywhere is paralel to the bottom (i.e. wave lengths are large compared with water depths)

Discretization – Branches

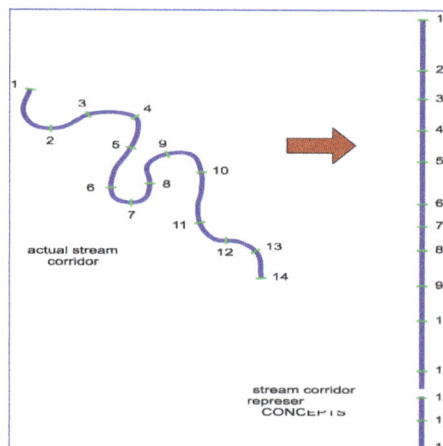

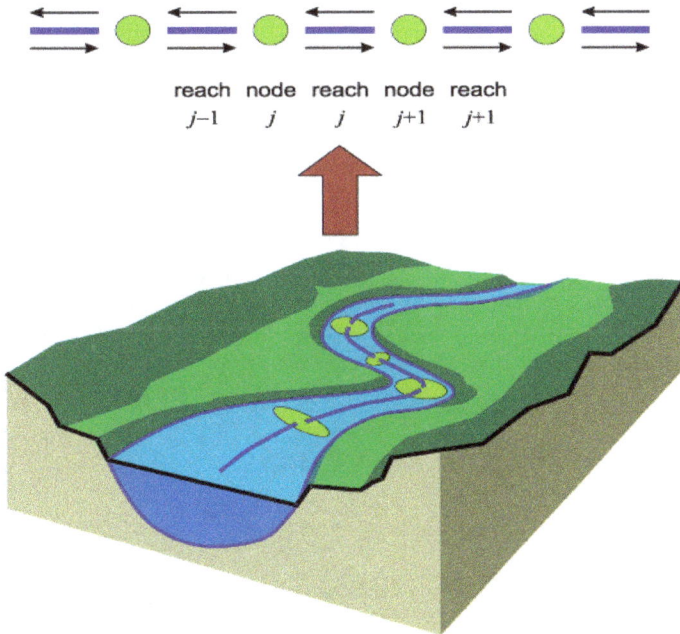

Discretization - Cross Sections

It is required at representative locations throughout the branches of the river and must accurately represent the flow changes, bed slope, shape, flow resistance characteristics

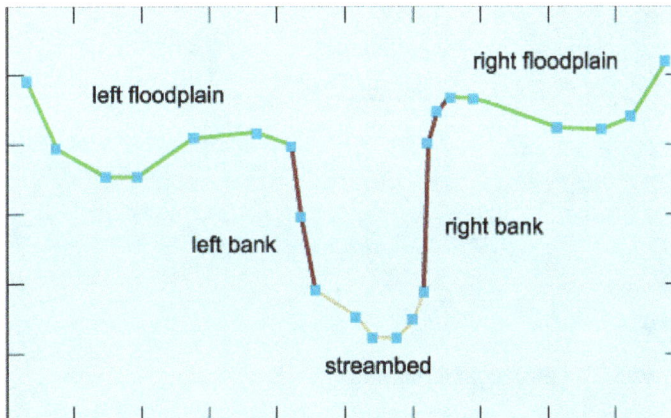

Friction formulas are of two types:

- Chezy
- Manning

For each section a curve is made with wetted area, conveyance factor, hydraulic radius as a function of water level.

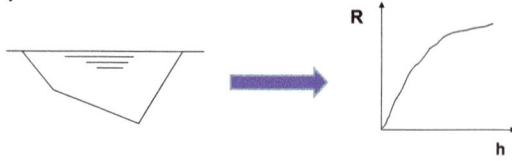

Typical Model Schematisation

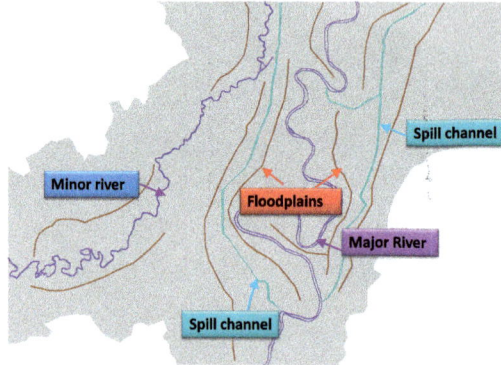

Mike 11 Main Menu

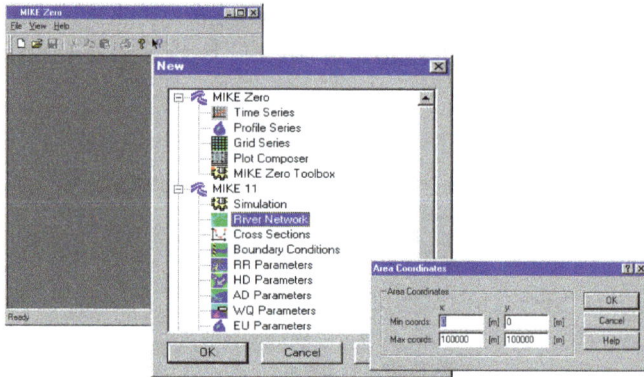

Extraction from DEM

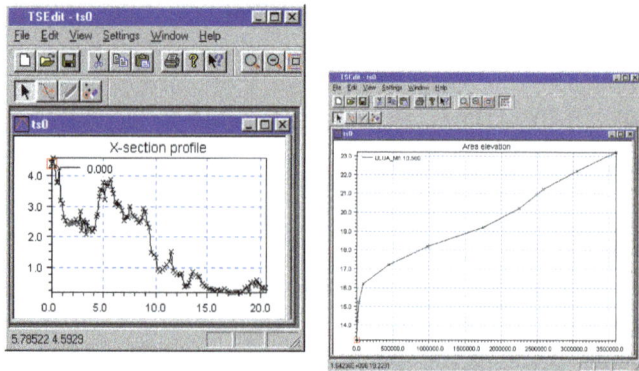

Import to MIKE 11

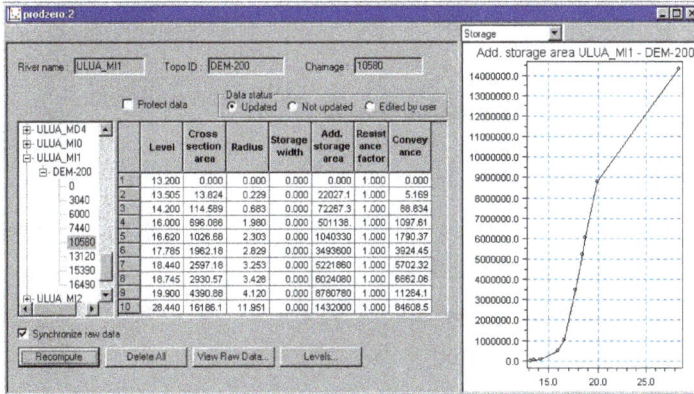

Menus and Input Files Editors

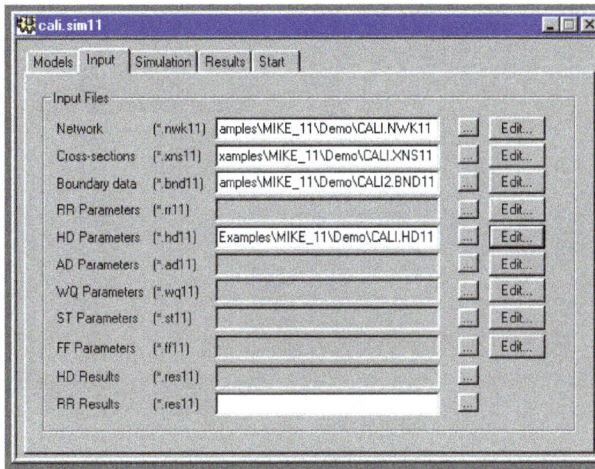

Network Editor

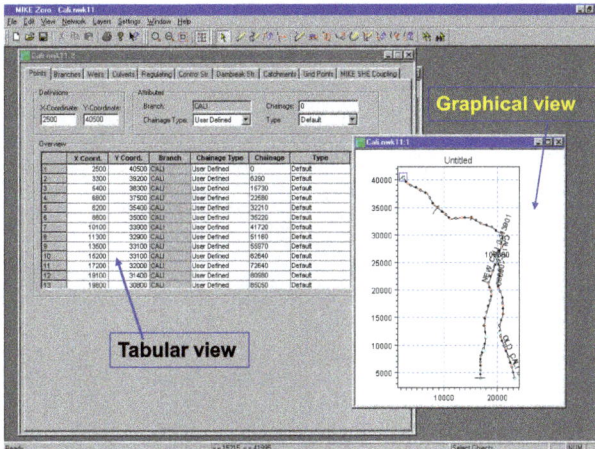

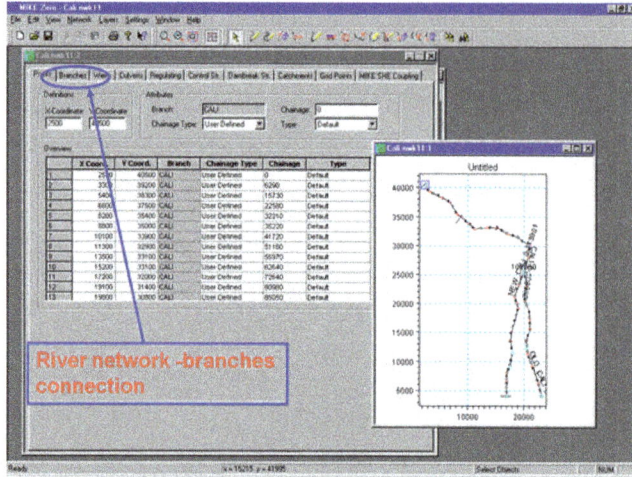

River network -branches connection

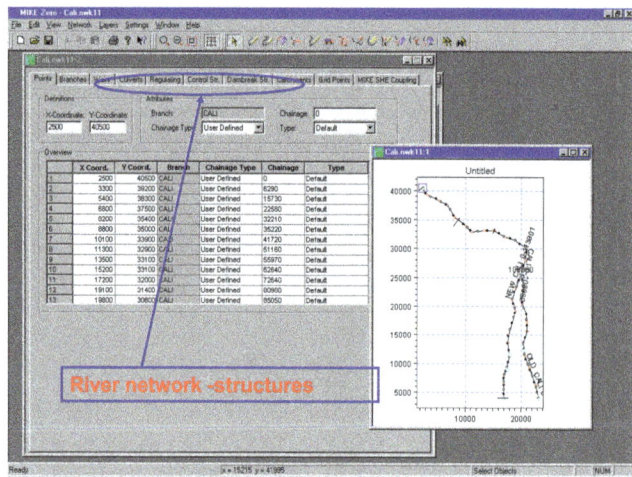

River network -structures

Parameter File Editor

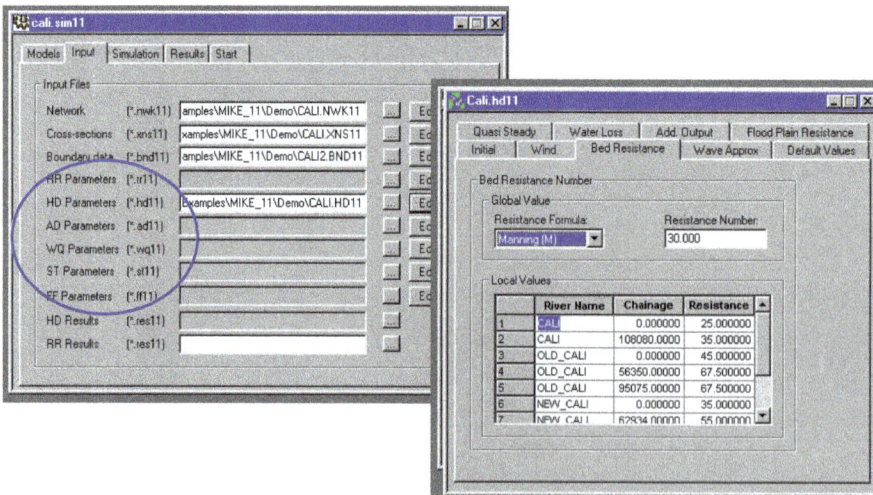

Parameter File - Coefficients

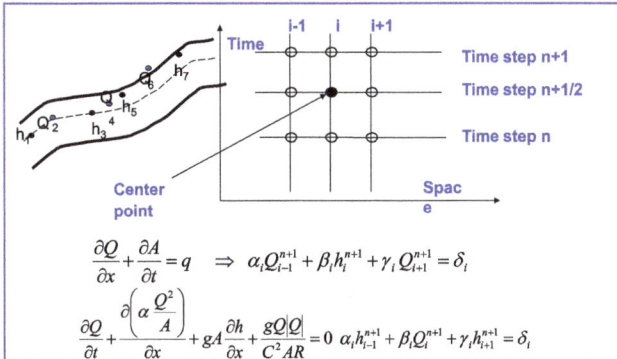

Limitations of MIKE 11

Because of its numerical limitations, MIKE 11 cannot model the supercritical flow downstream of the weir.

- For the low-flow case, the downstream water level is over-estimated by a factor of 8 .This high tailwater, impacts on the flow conditions on the weir, causing a significant error in the upstream water level.

- The incorrect tailwater has less impact for the high-flow case. There is still significant error in the predictions across the weir, but the upstream water level is almost correct.

MIKE SHE

It is an advanced integrated hydrological modeling system that simulates water flow in the entire land based phase of the hydrological cycle rainfall to river flow, via various flow processes such as

- overland flow,

- infiltration into soils,

- evapotranspiration from vegetation, and

- groundwater flow.

MIKE SHE Features

1. Integrated:

Fully dynamic exchange of water between all major hydrological components is included, e.g. surface water, soil water and groundwater.

2. Physically based:

It solves basic equations governing the major flow processes within the study area.

3. Fully distributed:

The spatial and temporal variation of meteorological, hydrological, geological and hydrogeological data.

4. Modular:

The modular architecture allows user only to focus on the processes, which are important for the study.

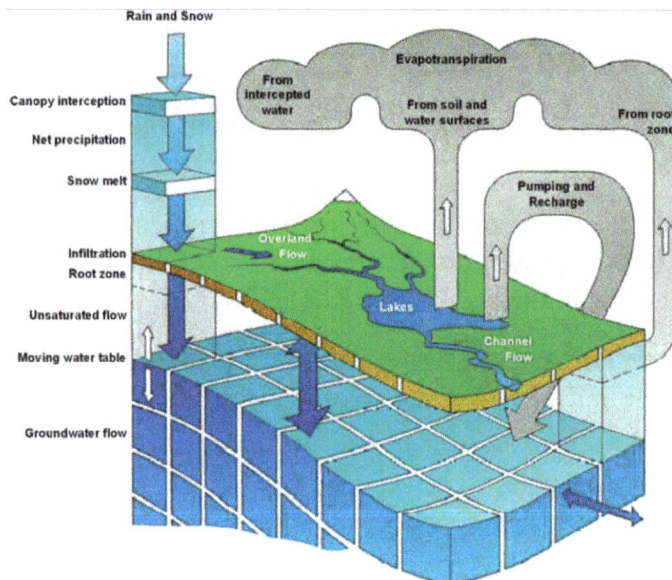

Hydrological Processes simulated by MIKE SHE

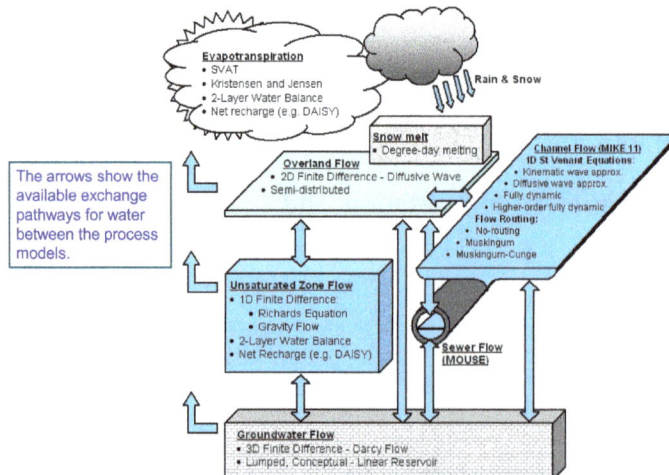

Schematic view of the process in MIKE SHE, including the available numeric engines for each process.

Urban Flood Risk Mapping using MOUSE, MIKE 21 and MIKE FLOOD

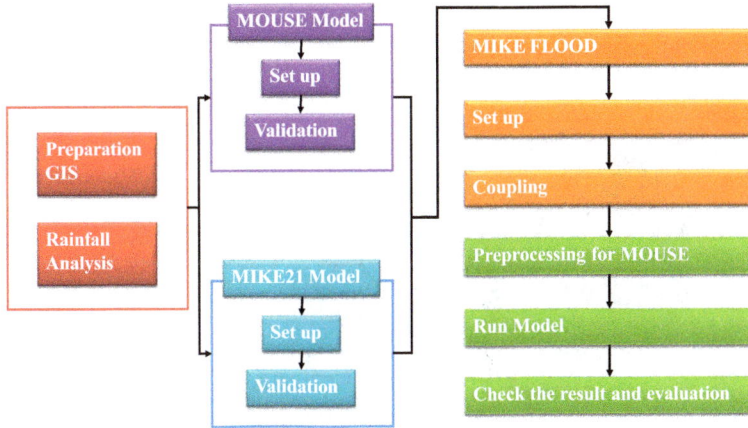

Urban Drainage Network

- Import of drainage network into MOUSE

- Setting up Urban Drainage model with MOUSE

- Validation

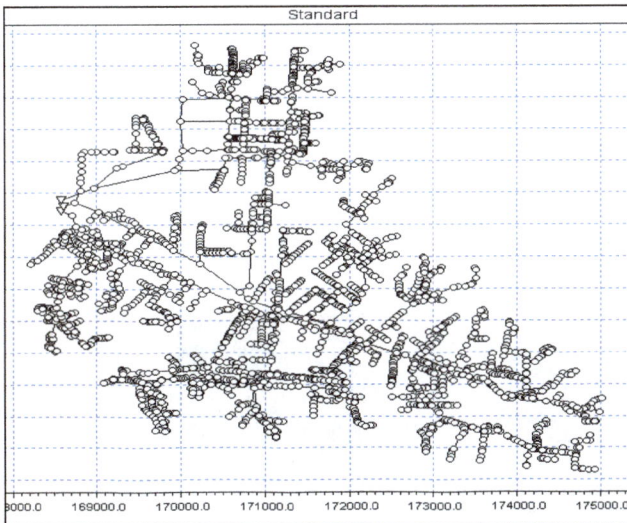

Urban Drainage Modelling

- Import Ascii data-set as bathymetry

- Setting up Urban Bathymetry with MIKE21

- Validation

Option	Value
Time Step	1sec Max Cr=0.4
Grid Size	10m

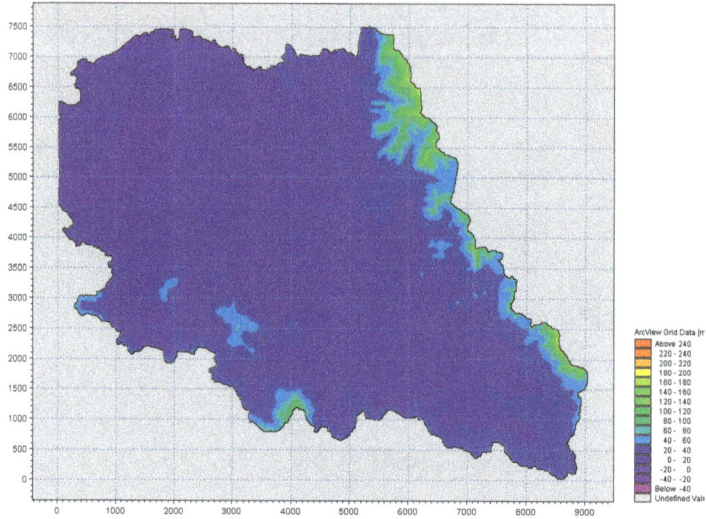

MIKE FLOOD Modeling

- Setting up MIKE Flood model

- Coupling : Link MOUSE Manholes to MIKE21

- Preprocessing for MOUSE Model

- Running Model

- Check the results

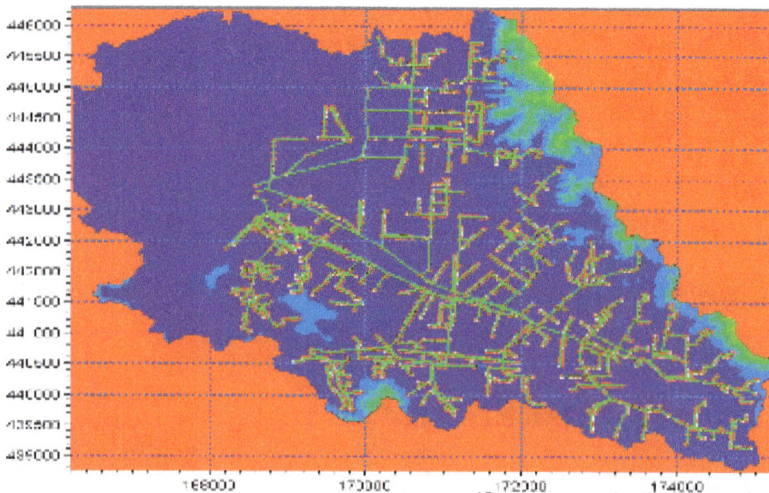

Results Analysis

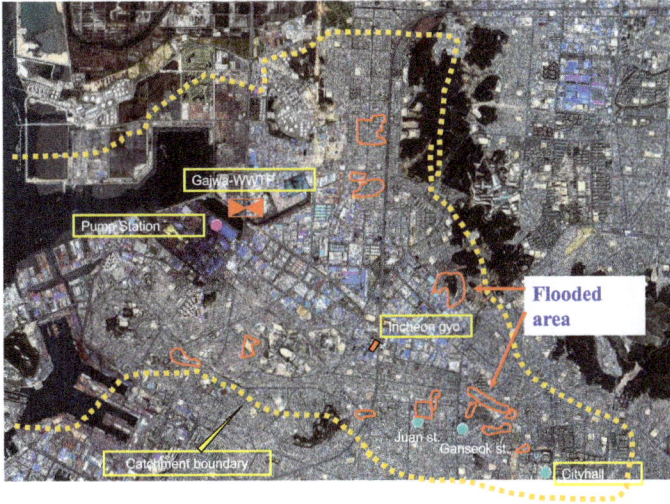

Comparing the Result (Modeled VS Reported)

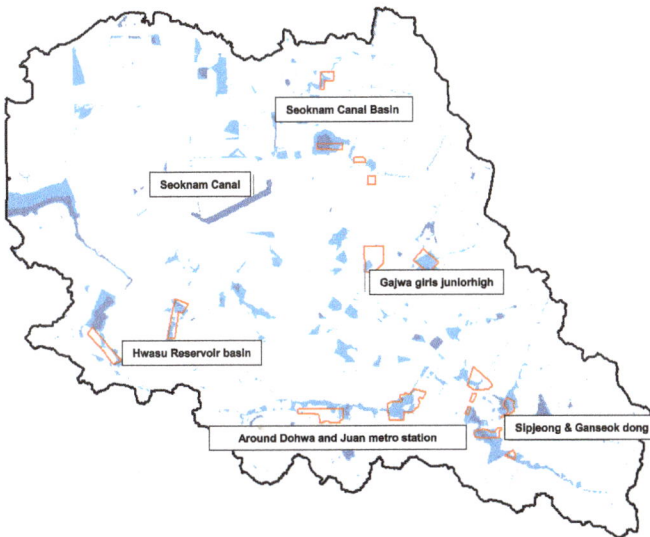

Permissions

All chapters in this book are published with permission under the Creative Commons Attribution Share Alike License or equivalent. Every chapter published in this book has been scrutinized by our experts. Their significance has been extensively debated. The topics covered herein carry significant information for a comprehensive understanding. They may even be implemented as practical applications or may be referred to as a beginning point for further studies.

We would like to thank the editorial team for lending their expertise to make the book truly unique. They have played a crucial role in the development of this book. Without their invaluable contributions this book wouldn't have been possible. They have made vital efforts to compile up to date information on the varied aspects of this subject to make this book a valuable addition to the collection of many professionals and students.

This book was conceptualized with the vision of imparting up-to-date and integrated information in this field. To ensure the same, a matchless editorial board was set up. Every individual on the board went through rigorous rounds of assessment to prove their worth. After which they invested a large part of their time researching and compiling the most relevant data for our readers.

The editorial board has been involved in producing this book since its inception. They have spent rigorous hours researching and exploring the diverse topics which have resulted in the successful publishing of this book. They have passed on their knowledge of decades through this book. To expedite this challenging task, the publisher supported the team at every step. A small team of assistant editors was also appointed to further simplify the editing procedure and attain best results for the readers.

Apart from the editorial board, the designing team has also invested a significant amount of their time in understanding the subject and creating the most relevant covers. They scrutinized every image to scout for the most suitable representation of the subject and create an appropriate cover for the book.

The publishing team has been an ardent support to the editorial, designing and production team. Their endless efforts to recruit the best for this project, has resulted in the accomplishment of this book. They are a veteran in the field of academics and their pool of knowledge is as vast as their experience in printing. Their expertise and guidance has proved useful at every step. Their uncompromising quality standards have made this book an exceptional effort. Their encouragement from time to time has been an inspiration for everyone.

The publisher and the editorial board hope that this book will prove to be a valuable piece of knowledge for students, practitioners and scholars across the globe.

Index

www.ingramcontent.com/pod-product-compliance
Lightning Source LLC
Chambersburg PA
CBHW061950190326

41458CB00009B/2839